Shapes in the

What Cloud Computing shapes are you looking for?

KAUAI CLOUDS AT SUNSET

A STRAIGHTFORWARD BUSINESS-DRIVEN PERSPECTIVE ON
CLOUD COMPUTING CHALLENGES AND OPPORTUNITIES
FOR THE CURIOUS AND INNOVATIVE AMONG US

Chuck Nelson

Book cover design by Shelley Neves

First printing: February 2011

Paperback ISBN-13: 978-0-9833709-0-1 (0-9833709-0-7)

Published in Gilroy, California, USA

E-Book ISBN-13: 978-0-9833709-1-8 (0-9833709-1-5)

Library of Congress Control Number: 2011902630

Trademarks

Warning and Disclaimer

Thank you to my wife, Robin, to my mother, Ruth, and to my children Nicole and Erik, for your unconditional love, support, and encouragement.

****

In loving memory of Burnell and Maureen.

Author Profile

Chuck Nelson is a senior product management professional with a proven track record of identifying opportunities and delivering innovative solutions that transform business challenges into strengths, competencies, and profits. He enjoys working with creative highly-motivated teams that are intent on delivering great solutions for advances in quality of work and life.

Chuck has a unique perspective on the integration of IT and business management, having worked for more than 25 years in technology companies on major initiatives as the solution designer and project manager for internal projects, as a senior product manager to address substantial existing gaps, and as a director of product management, strategy and planning to deliver extensive new products and features.

Chuck's products, services, and solutions have generated outstanding product lifecycle revenue, profit, and market share results for startups, medium-sized, and large companies, including HP, Apple, Dell, and SAP. He is a certified market-driven product manager and a certified Scrum Master, with multiple patented innovations at SAP. Chuck has defined and implemented extensive business processes while leading global teams to achieve best-practice performance, with domain expertise in guided processes, CRM, sales and marketing, demand planning, channel management, SCM, analytics, accounting and finance functions.

Chuck's most recent endeavors encompass a variety of interests around Cloud Computing solutions and IT Management. He has provided well-received product strategy presentations to SaaS ERP solutions and to a SaaS education eBook startup. He has also defined an innovative design concept for a Cloud-based professional opportunities network with a 3-way ecosystem, encompassing integrated applications for the job-seeking and career-planning individual, for the 3rd party recruitment broker, and for the hiring enterprise. Chuck's latest business venture was an extensive SaaS-based application, platform, and collaboration ecosystem for resource Sustainability Management and Optimization, again applying his great interest in business solutions, Cloud Computing and IT Management. He continues to look for opportunities in these areas of interest.

Foreword

I have always been intrigued by new ideas and opportunities. When I look at something that is working very well and admirably serving its users, I tend to carefully examine the solution in order to learn from its creators and their experiences that led them to design and implement the solution the way that they have. If it is something of great interest to me, I invest time and energy in taking the solution forward to the next level(s).This behavior is usually a great blessing and occasionally a small curse.

When I first encountered any mention of Cloud Computing, it was presented in extraordinarily vague and flowery terms, where some colleagues were talking excitedly about great new opportunities while others laughed and said it was really nothing new, just a lot of hype. The next encounter I had on the subject left me feeling teased, because everyone involved in the discussion indicated that the subject was huge, had a lot of moving parts that seemed to be constantly changing to protect the guilty, and that it might take 5 years for it to shake out. Most follow-on discussions ended with "It's the next big thing in business solutions". Well, it was a small lure, but I was indeed hooked.

The problem that I had was the old blind men and the elephant routine. You know, where one guy touching the elephant's leg is thinking he has something different from the one touching the elephant's trunk. And the guy at one "end" of my Cloud journey could only speak technology and not understand the business drivers, while the guy at the other end wanted the nicely-packaged Cliff notes on all things Cloud, and neither were terribly interested in rational research of what it is (perhaps because their research might change their foundational perspective and make them have to start all over again).

So I found that the more I thought about it, the more questions that I raised and tried to answer. I could never find information that looked at the Cloud quite the way that I was. And as I began to structure and compile my conclusions and meditations about the Cloud, I found that I was probably looking at it differently. Yeah, I know, just another guy holding onto some appendage of Cloud Computing, right?

Well, maybe.

But I wanted to know what the world thought of when it heard Cloud Computing. What happened in the history of business technology solutions to lead us to this new paradigm? There has to be reason in there somewhere, something that will make business more profitable, reduce risk, etc. What were the benefits from Cloud Computing? What were the limiting or inhibiting factors? What were its parts? How did they fit together? What existing descriptive methods or prescriptive solutions would become key components to moving forward to a broad basis of Cloud Computing adoption? What was all of this granularity-in-metrics and subscription pricing about? What were some of the use patterns and use cases that differed based on whether you were a service provider or a service consumer, or on the type of Cloud service, or the Cloud deployment model? What if you are moving from an on-premise owned-everything situation to an on-demand pay-as-you-use-it situation? How should one evaluate this?

Maybe there are others that are curious just like me. So I wrote this book, and here we go.

I hope you will enjoy the book,

Chuck Nelson

Principal, Nelson Consulting

shapesinthecloud@gmail.com

A Roadmap for the Book

"Shapes in the Cloud" provides a business and IT management perspective on applying Cloud Computing technology to multiple enterprise situations. It reflects extensive experience in the definition and realization of products, services, and solutions that serve business goals and objectives.

A roadmap of the book is provided here to assist the reader's sense of where they are and where they are going as they read through the book.

- Chapter 3 introduces the key **players in the Cloud** – Providers, Consumers, and Users – and provides a comprehensive definition of Cloud Computing. It explains the evolution of business drivers that brought us to a Cloud era, and the benefits and inhibiting factors around cloud Computing.

- Chapters 4, 5, and 6 provide **business management insights** into some of the technologies and architectural foundations of Cloud Computing. **Virtualization** of IT resources, **multi-tenant** software, the various Cloud **deployment models**, and a **framework** for understanding how all the parts work together, are introduced.

- Chapters 7 through 13 take the reader through the Cloud Computing **framework's layers and components**, focusing first on defining and evaluating **general Cloud readiness**, then doing the same for the **infrastructure, platform, application, security, and Cloud management** areas of interest. Some existing providers' offerings are identified in general and in context.

- Chapters 14 and 15 focus on **ITIL, ITSM, BSM, and Cloud Management** as significant components in the **IT Management** challenges associated with managing services throughout their lifecycle, culminating in a brainstorming discussion on **integrated cloud lifecycle management systems and solutions**.

- Chapter 16 provides an introduction into the **metrics and granularity** possible in the **types of charges** that providers and consumers will encounter in their Clouds.

- Chapter 17 is a comprehensive discussion on **consumer usage patterns and use cases** from startups to mature businesses. A specific use case for **software providers** is included here, because Cloud Computing is a perfect fit for their **operations**.

- Chapter 18 provides a focus on the **Cloud offerings of software providers**. Cloud Computing represents a very unique business model for all subscription-based providers, especially so for the software providers. Cloud Computing platforms and applications can be great enablers for startups but a bit more complex for existing on-premise software solution providers that are moving towards Cloud offerings.

- Chapter 19 concludes the book by providing some insights into how the **Cloud Computing promise of agility** is accomplished through the use of agile software development methods, specifically through a **comparison of Scrum to the Waterfall** method of development.

Outline of Contents

2

So Many Shapes Are Possible

An Introduction

So Many Shapes Are Possible

I can recall more than one occasion when I found myself gazing at a variety of clouds moving across the sky and losing myself for awhile as I was matching the "look" of people, places or things to the appearance and shape of individual clouds. One cloud looked like the face of Abraham Lincoln, another like Mount Rushmore, while trees, rivers, airplanes, cars, and birds also appeared. It was almost like imagining just about any shape or thing was possible. I just had to open my mind to the possibilities and then I could see many options for each shape. Sometimes the shape would actually change as the wind moved the cloud across my view, allowing me to imagine other matching options.

When I was thinking of the many things that I find so interesting about Cloud Computing, I thought of how applicable the analogy is to these shape experiences, and that most people probably have done something like this cloud-gazing at some point in their lives. It occurred to me that Cloud Computing is many things to many people, and the differences in what people think seems largely to be based on their challenges, where they have been, where they are, and where they think they might be going.

A small business that is just starting up, with very modest capital available for non-recurring and recurring costs of operations, may see the Cloud as enablement with reduced barriers to profitability. A more mature business may sense the opportunity to reduce its operating costs while focusing more of its resources on its core competencies and consistently delivering on its value proposition. A large or very large business may recognize tremendous opportunities for increased agility and reduced time-to-market where new divisions or business units are spawned by innovation, merger, or acquisition. All may also feel varying degrees of "analysis paralysis" as they contemplate how best to consider moving one or more current business software platforms and applications towards on-demand services. This is a compelling inertial challenge, as they know quite well what they have in their current landscape, but the competitive pressures in existing markets call for improved profitability while opportunities for extension or expansion to new markets are increasingly demanding greater agility.

On the other hand, for the various technology companies that offer and/or enable optimal Cloud services in the Cloud Computing solutions value chain, Cloud Computing represents the possibility for a next great wave of opportunities for innovation and growth. Using Cloud services, the startups and their investors have substantially reduced their scope of initial investment and risk of loss while maintaining or improving their probability for success. The existing companies might recognize how their technology products, services, and solutions will provide one or more values that partners or customers will pay for in order to resolve a business challenge. This is especially challenging for software suppliers who must complement their existing on-premise business model with an on-demand business model.

This book provides a general introduction to the considerations, challenges and opportunities associated with Cloud Computing. It presents the various perspectives of Cloud service providers and Cloud service consumers, with the objective of informing all parties regarding these different perspectives.

3

A Cloud By Any Other Name

Players, Drivers, Factors

Discussion Areas in This Chapter

- *Cloud Computing definitions*

- *The business evolution to the Cloud*

- *Positive attributes of the Cloud*

- *Limiting factors for Cloud adoption*

A Cloud by Any Other Name

There is a tremendous amount of discussion and interest regarding Cloud Computing. Let's start by establishing some basic attributes around Cloud Computing and this will provide context for the rest of our discussions.

A. Cloud Computing Definition

There are several parts to the definition of Cloud Computing, some existing before the advent of Cloud Computing. This section will establish the basic parts and how they relate to one another and to Cloud Computing in general.

1. IT Capabilities Delivered as Services

The management and contributions of Information Technology (IT) have gone through a number of evolutionary changes even before the introduction of Cloud Computing solutions. Not the least of these pre-Cloud changes is the relatively recent formal recognition of the following:

- A service is a means of delivering value to one or more customers by facilitating outcomes that these customers want to achieve without the costs and risks that come with ownership of the solution components.

- Service Management is a set of specialized organizational capabilities for providing value to customers in the form of services.

- A business process is an activity that is owned and carried out by a business to contribute to delivery of a product or service to its business customer.

- An IT service is a service based on a combination of people, processes and information technology resources that is provided to a customer within a business to directly or indirectly (through other IT services) support the customer's business process.

- IT service management is the definition, implementation and management of quality IT services that accurately reflect the alignment of IT resources with support for business objectives and priorities to meet the needs of the business.

Figure 3-1 below provides a basic presentation of IT services supporting the business to achieve success. An enterprise's applications, platforms and infrastructure resources are redefined and packaged as business services, supporting IT services, and IT infrastructure services that align IT and business objectives and priorities.

It should be noted, however, that managing and delivering IT resources as services aligned with business objectives and priorities is not a hard prerequisite for an entity providing or consuming Cloud Services. It is a pre-existing paradigm in more mature IT organizations in most medium-to-large size companies.

IT Services Support Business Processes

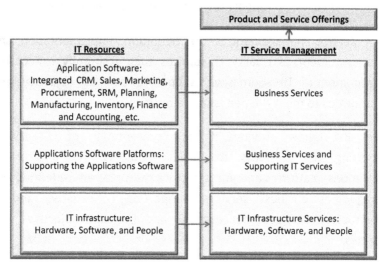

Figure 3-1: IT services support the business objectives and priorities.

2. Players in the Clouds

There are three fundamental participant roles in Cloud Computing scenarios: the provider, the consumer, and the user.

- The supplier of the service is the entity that is commonly referred to as the "Service Provider". The service provider is an organization that is supplying services to one or more consumers. It manages the costs and risks of the services, instead of the service consumer, spreading these costs and risks over multiple consumers whenever possible. An internal IT service provider provides IT services to a consumer within the same business entity. An external IT service provider provides IT services to a consumer that is a separate business entity.

- The "Consumer" is the organizational entity that has the business relationship with the service provider. It may be an enterprise, a government agency, or an individual. Enterprise and government entities are the primary focus of this book, as they consume services to help them deliver business products or services to their consumers. In contrast, individual consumers are not businesses, they consume the services for their own personal value reasons, and they do not manage IT services in general.

- The "User" leverages the service acquired, configured, customized, implemented, and/or accessed by the consumer. The actual end-user is consuming the service. Users may be hosted anywhere and might be a human, an organization, or a machine. With individual consumers, the consumer and the user are the same individual. With enterprise and government users, the user is typically one of several potential users within a group or function. The user may also be an integrated system that is using the service as part of an end-to-end business process.

3. Cloud Computing Definition

There are many different ways to define Cloud Computing, just as there are many shapes in the clouds. For the purposes of this book, some Cloud Computing solutions may not have all of the following definition attributes, but the most complete instances of Cloud Computing solutions will meet all of these criteria.

Cloud Computing is a computing solution model employing internet technologies for the scalable and elastic provision, application, and consumption of computing resources that are pooled, shared and delivered on-demand as metered subscription-based services available to one or multiple consuming entities over a network (local area, wide area, Internet). A computing solution service is provided as it is required and, if the service isn't free, the consuming entity is charged by the service provider as the service is consumed. The identity of the physical resources used in the performance of the service is an abstraction to the consuming entity and is known only to the provider. Some expressions used to describe this model include "ready-to-go", "pay for what you use when you use it", and "pay-as-you-go".

If we break this definition down into an outline form, these are the critical parameters of Cloud Computing:

- **Pooled and shared resources**

  Computing solution resources may include compute processing, memory, storage, network bandwidth, operating systems, application platforms, applications, etc. The service provider's resources are pooled to serve one or multiple consumers with physical resources abstracted from the virtual resources that are dynamically assigned to meet consumer demand.

- **Internet technologies over networks**

  Browsers, web servers, application servers, client/server, SOAP, REST, TCP/IP, MVC, Flash, Java, Ruby, LAMP web application framework, and other web technologies are the basis for Cloud Computing solutions delivered over networks. Resources are accessed through standard solutions using heterogeneous thin or thick-client platforms (mobile phones, laptops, and PDAs).

- **Automated on-demand self-service provisioning**

  An authenticated subscribing consumer can unilaterally and automatically provision computing resources according to pre-agreed service level agreements as needed without requiring human interaction with each service's provider.

- **Metered, measured services**

  Services can be measured and charged because a metering approach exists at one or more levels of each resource's abstraction that fits the service provided – stored terabytes, I/O transactions, compute hours, monthly access and support, etc. Resource usage is monitored, controlled, reported, and billed with common transparency for the provider and consumer.

- **Scalability with rapid elasticity**

  Resources are highly scalable, with rapid elasticity to quickly scale-out and scale-in to the required provision levels. A consumer should perceive unlimited resource availability that it can demand in any quantity or volume at any time.

The preceding (and currently) dominant model of providing and applying computing resources is generally referred to as the on-premise model. In its purest form, it requires each prospective resource consumer to buy, own and maintain their needed facilities, hardware, and software in advance of their use. In the most common example of this model, computing networks may be local or wide area, internal or the Internet, but the consumer directly owns its datacenters, servers, software, supporting networks, and employs its own IT staff to customize and manage all of this to fit their unique needs.

The contrast between on-premise and on-demand models of provisioning and application of computing resources is significant. In a world where nothing would ever change, the on-premise model might work well. But in our world, agility is critical to driving and applying technology because change is constant. Perhaps the purest form of agility might be to buy and own as little as possible while using the services provided by subject matter experts to accomplish your objectives.

Renting a furnished home with utilities, gardening, housekeeping and maintenance all included for a fixed monthly rate is perhaps an analogy of a common on-demand resource provision. From the Cloud perspective, some well-known examples include Google's GMAIL, WebEx's web-based conferencing, eBay's auctions, salesforce.com's CRM, Ariba's sourcing management, NetSuite's ERP, BMC's on-demand ITSM, and Amazon's web services for Cloud infrastructure platforms and solutions.

B. An Evolution

An evolution in business practices over the past 50 years has been enabled by very innovative technologies. These are some of the primary drivers that have led to the introduction of Cloud Computing and the increasingly rapid growth in demand for and adoption of Cloud Computing solutions.

1. Business automation enables growth and expansion

Limited technologies were effectively restraining economic growth and expansion. Manual processes had been refined up to the limits of the technologies available, primarily carbon-copy paper, pencils, and warehouses filled with hardcopy storage. Successful results came predominantly from highly-experienced people who had become skilled in their respective work through many of years of repetition. Best business practices reflected modest empirical improvements and "tribal knowledge" that was passed down through generations of workers. Business processes needed to be automated.

2. Automation was limited

Larger companies with sufficient financial, technology, and people capital began to invest internally in IT staffs and large computer centers to automate business processes. Business analysts were used to define the requirements for these processes, and each business began to develop and implement its very own unique automated business solutions predominantly using mainframe and mini-computers. These in-house solutions were very expensive, requiring substantial company resources and a great deal of time, effort, and energy to realize.

3. Automation was proprietary

Automation typically produced substantial improvements in quality, capacities, costs, timely and accurate decision-supporting data, etc. Each company defined its own solutions in its own ways. Therefore, companies perceived their computing resources and their legacy systems as closely-guarded and strategic intellectual property assets that produced differentiable values and competitive advantage. The playing field for less wealthy companies was anything but level.

4. Process improvement was painful

Inflexible systems were developed for individual functions and were not very well integrated with one another, leading to siloed isolation. As new ideas or approaches for procedures, processes, and technologies became available, new levels of investment and re-engineering were required, often where each succeeding version of in-house development updates might be more expensive than the last one. The enterprise was heavily invested as a technology company in addition to serving its primary markets.

5. Focus on core business competencies

Best practices and process standards for whole industries became increasingly well-defined. Software companies began to form around these standards and practices with sufficient technology to create solutions that could be packaged and sold as products to all companies. The legacy-driven companies began to evaluate the cost and value of in-house process development. In a strategic context, the companies realized that buying the software, hardware, and support for their business processes from entities whose core competencies were ongoing software and technology development would still meet or exceed their quality, volume, cost, and agility objectives. Any unique features that they might sacrifice probably did not represent as much competitive advantage as always using the latest technologies from an external vendor at less cost.

6. Front-end costs were barriers to growth and expansion

Any startup, new division or business unit would require IT staff to setup and run the new computing solutions. Expected patterns of business and demand cycles would be forecast, and new hardware and software would be purchased and implemented as costs to initialize the enterprise for business. These initial investments in computing resources were often driven by capacity requirements for peak demand processes, and might be equal to or greater than the investment in product development, sales, or marketing functions, thereby perhaps reducing the opportunities for growth or expansion.

7. Locked-in by substantial capital expenditures

Some global companies have spent billions of dollars to commit to each new landscape of solutions. Their costs would include buying the software licenses and consulting services for redefining their processes, customizing the software, creating custom EDI solutions and batch programs for integration, training staff, documentation, etc. They would also pay maintenance fees to their software vendors that might range between 15 and 25% (of the associated license fees) per year for support and ongoing improvement. In this manner, the software vendors have non-recurring and recurring streams of substantial revenue that locked-in their installed base.

8. Agility for new business

Internet technologies have evolved to fully enable bidirectional processes. These include providing a new easily-accessible channel; substantially growing a business while eliminating most of its "brick-

and-mortar" business model costs; and creating new businesses that exclusively employ the web as the only channel for their business. Great web-based concepts, vision, and entrepreneurial management have produced exceptional innovations in web-enabling tools and applications. Network, database, and storage companies grew exponentially in parallel with server and computing technologies to support the business needs. Software companies were forced to help their customers consolidate software instances, while a variety of technology standards and web application frameworks began to appear that were capable of producing increasingly robust platforms for developing and integrating web-based enterprise business applications.

9. Costs of ownership are too high

The technologies that made all of these tremendous changes in business possible have concurrently produced dramatic increases in computing capacity at incredibly reduced costs. But the reductions in cost are still not enough to overcome the drag on businesses. New demands drove investments in faster-time-to-market, world-class product and business development, quality and customer service, transparency and accountability, IT governance and business compliance, and ease-of-use with security, and these demands have all initially increased the cost of doing business.

Each data center must contend with acquiring and managing office space, power, cooling, bandwidth, networks, servers, and storage and a team of experts to install, configure, and run them. Each of the systems used by the consumer will need development, testing, staging, production, and failover (means for ensuring high availability of some critical resource) environments. And this level of demand may actually be multiplied by the number of mission critical solutions, across dozens or hundreds of applications. So it's easy to see why the biggest companies with the best IT departments weren't getting the applications they need. Small and medium businesses didn't stand a chance.

The cost of owning the entire computing solution could easily become a huge business disadvantage.

10. On-demand services

Some companies became exceptional in their development of web solutions and realized they had developed core competencies in the creation of web applications, platforms and infrastructure. They recognized that every other company that was interested in web solutions would be required to develop these competencies as well. Instead of seeing their internal technologies as IP to safeguard, they recognized the opportunity to sell their solutions as services to other companies that would be willing to pay for it on an as-needed basis. They were right because the demand from startups, SMB (small and medium businesses), and large companies for the agility, quality, reduced risks and costs, etc., in computing solutions is definitely there.

Some examples of this include:

- Google search engine and Yahoo commerce platforms led to free email services.

- Salesforce.com created a very robust CRM suite of applications that became the poster child for making the case that software-as-a-service deliverables could be successful.

- Salesforce.com recognized the power of the platform it had developed as an enabler of other SaaS innovators to quickly get to market with their products with reduced costs, and then offered Force.com.

- Amazon used its investment in infrastructure and platforms that supported its eCommerce business to offer its AWS (Amazon Web Services) to enable other companies to quickly and easily afford the creation of their own platform and application solutions.

C. Defining the Positive Attributes of the Cloud

There are many potential business drivers and related benefits associated with the adoption of Cloud Computing solutions. An outline is provided below of the attributes of Cloud Computing addressed in the following section for information about the drivers and benefits of Cloud Computing.

> Pay-as-you-go
>
> Providers operate with greatest economies of scale
>
> Elasticity
>
> No front-end costs
>
> Focus on core competencies
>
> Fixed costs become variable costs
>
> Capital expenditures become operating expenditures
>
> Resources are abstract and undifferentiated, with associative costs
>
> The consumer's business is much more agile

1. Pay-as-you-go for metered usage

The Cloud model enables the provision of computing resources to the consumer as individual, separately allocated and atomically-defined building blocks available for their solution. The consumer defines the desired levels of service, seeks the best price for the services, and then pays according to these predetermined subscription rates as providers allocate and deliver resources responding to the consumer's demands. The consumer can choose to consume none, one, many or all of the services offered and available from the Cloud service provider.

2. Providers are SME's with greatest economies of scale

A service provider must become a SME (subject matter expert) for the scope of its offering. It will sustain its business if it can continue to make an acceptable profit through the subscription business model. It is likely to improve that profit level even more if it can grow its business by consistently providing an increasingly attractive spectrum of quality services that add great value to delighted customers for competitive prices. The larger the customer base, the more effective the prioritization for continuous improvement in both technology and operations. Basically, if a solution is honed to satisfy a greater variety of situations and applied far more often than most, a provider should be able to recognize cost reductions due to the sheer volume it supports. This volume means that the provider can buy hardware, software, real estate, power, bandwidth, etc. at the lowest possible prices, and staff the best skills available for development, engineering, and/or operations at global scale levels. While extremely large global companies might come closest to affording these economies of scale in on-premise models, the Cloud effectively makes this leverage available to even the smallest business.

3. Elasticity and scalability

One of the most challenging aspects of business is the unpredictable nature of demand. Demands can vary widely depending on many factors, not the least being that you just might be wildly successful in your venture. Some also refer to this demand volatility as "spikey" or "bursty".

Now the consumer can afford success and sustain acceptable levels of customer satisfaction. One key defining attribute of Cloud Computing is the fact that it enables the consumer to avoid advance provisioning as this is now the problem of the provider. Each consumer can "expect" the Cloud to provide nearly-infinite scalability. It is the service provider that now provides economical and efficient scaling up and down of resources in timely response to demand changes, often in seconds or minutes versus months for on-premise changes. New hardware can be brought online very quickly to deal with unexpected increases in demand from large internal jobs or external web traffic. And these resources are then returned to the Cloud's open capacity when it is no longer needed to meet demand.

4. No front-end costs

In many ways, Cloud Computing is a "leveler of the playing field". The on-demand nature of the Cloud means that you don't have to buy resources to support your expected peak demand levels before you even take your first service request. And the Cloud model provides for the temporary allocation of resources for any idea or experiment the consumer might pursue. Those huge front-end investment barriers that used to stifle many innovative ventures before they even had a chance have been removed.

5. Focus on core competencies

This applies to both the service provider and the service consumer.

The service provider's focus is on its core competencies in developing the best application, platform, infrastructure, network, and implementation and management tools. It understands the success vectors present in technology and market trends. In this context, it is probably providing the Cloud Computing solutions for public consumption, so it must focus on supplying services that are applicable, reliable and optimally value-priced. It will deliver the best solutions available for the consumer's situation and can even become a critical partner in the consumer's ability to execute, lead and differentiate itself from its competition.

In contrast, the consumer's competencies reflect its knowledge of its business opportunities for its offerings across its own targeted markets. The Cloud model means they can concentrate on (and invest in) the things that will make their products excel in whatever industry focus they have. This is an incredibly important aspect of the Cloud, because it intrinsically moves to align all of the strategies, planning, and operational execution of businesses, while securing the best possible ongoing computing support for the consumer's business activities from external expert sources.

6. Fixed costs become variable costs

One of the greatest challenges in any business is the ability to accurately understand the costs of doing business in a manner that allows making the right commitments at the right time in the right way. The cost of on-premise software can be millions in licenses and tens of millions in implementation-related expenses. Many of the costs associated with an on-premise data center are fixed in nature and not allocated or distributed based on actual demand usage. With the on-premise model, an enterprise is likely distributing the costs of a data center (including staff) that is predominantly idle at a 7% level of usage for 90 % of the time, allocating to its internal users based on some index that is not likely to accurately provide a high positive or negative correlation to actual business costs. In contrast, Cloud Computing has an expectation of exceptionally high utilization rates which should improve the integrity of its cost allocation.

The costs of Cloud Computing are by definition variable costs as they reflect the actual usage of the consumer's demands. The consumer pays for actual usage demand, and its commitments are to service levels and quality, sustaining elasticity without multi-year commitments to resources that become increasingly less competitive over time.

7. Capital expenditures become operating expenditures

Many of the non-recurring investments made at the front-end of any business venture startup or refresh are not only fixed costs, but they are very likely to be Capital Expenditures or CAPEX. Because they are viewed as longer-term investments that reflect a multi-year commitment, they become strategic anchors or foundation, they persist on the balance sheet of the business and are allocated over 2 or more years of operations according to GAAP (generally accepted accounting practices). The financial results therefore represent a compromised level of meaningfulness when used to project the effectiveness of the business management.

The Cloud Computing model eliminates most of the typical computing capital expenditures because the consumer doesn't make these initial capital investments, and its ongoing actual usage demand results in costs that it incurs as a part of its operations. They become demand-driven costs of the consumer doing business.

8. Resources are undifferentiated, with associative costs

The consumer does not concern itself with the operational challenges that the provider must overcome. The consumer cares only that the expected results are realized and the expected service levels are met in every instance per the terms of its service agreements with the provider. The provider's degree of difficulty in meeting the SLA is transparent to the consumer. The physical hardware and software instances used by a consumer are abstract and undifferentiated, meaning that they may well be different from one occasion to the next occasion for the same consumer, or different physical resources for another consumer's requests for the same exact functions.

On the other hand, the technical specifications of the technology stack used to support the consumer will change and improve as the provider sees opportunities to improve their ability to deliver expected service levels at reduced costs. The fact that the provider's resource costs have become associative also means that it is now much more capable of innovative solutions for large-scale partitioning challenges, e.g. when 10 servers for 1 hour are effectively the same cost as 1 server for 10 hours.

9. The consumer's business is much more agile

By employing the Cloud Computing model, the consumer doesn't have the burdens described above for initial investments and longer-term commitments that produce predominantly fixed CAPEX costs, and its business is now capable of much greater flexibility.

The consumer can now pursue those strategically critical initiatives that will enable its business to create new offerings, go to market on a timely basis, sustain its leadership, expand its offerings, penetrate new markets, increase its differentiation, etc., because its business planning is focused on the business and not burdened by forecasting the costs of required computing support. The consumer's risk management need not prepare extensive computing support plans to overcome oscillations in demand or create substantial upside investments in computing support when going after new opportunities with unknown demand patterns and unpredictable business drivers. The Cloud Computing model is perfect wherever businesses must respond quickly and efficiently to meet changing business demands.

D. Recognizing factors that inhibit Cloud adoption

There are several factors that seem to consistently get mentioned as the reasons to seriously consider the risks in adopting any Cloud solution.

- Security and privacy are the number one concern of any entity considering the Cloud. Most prospective Cloud Computing consumers agree that at some point the security of SME (subject matter expert) service providers may be better than anything the individual consumer might be able to produce, but that this point has not yet been reached. Security is discussed in detail later in the Cloud Framework section on Security. Basically, the concern is around:

  o Both public and private sector organizations' insider threats coming from "bad actors" or "co-opted workers" with unrecognized allegiances to external entities, due to money, threats, loyalties, etc. Requiring all workers that can affect their operations to be on premise, to permit direct observation, clearances, and protections are critical Cloud starter concerns for some institutions. This may restrict these entities to private clouds, limiting the spectrum of benefits that Cloud deployments would normally produce.

  o The integrity of internet business in general. Some cite the frequent major internet storms that have seriously compromised web commerce and operations.

  o The new challenges associated with managing data and automated business processes that are outside of the physical governance of the consumer's own IT resources.

  o The difficulty, complexity and exposure to corruption and other liabilities associated with the extensive application of virtualization and multi-tenant computing solutions in making the Cloud work. (Virtualization and multi-tenancy are introduced in the next chapter.)

o The strategic reliance upon multiple tiers of providers within the Cloud ecosystem.

- The investment that the consumer may have already made in on-premise data center, software license, and customization solutions, as well as desktop business computing solutions, is not easy to walk away from, especially when all current processes seem to be working well after a previous difficult transformation period. This "sunk cost" consideration is another inhibitor of Cloud migration, especially "big-bang" or "near-term", but enterprises should recognize that external forces from competitive sources, emerging market opportunities, and/or disruptive technologies have always affected and will continue to affect the efficacy of any non-Cloud application.

- The investment in the integration and implementation of new end-to-end business processes that will be required wherever legacy enterprise solutions must be integrated with the new Cloud infrastructure, platforms and applications. There is a deep abiding recollection of the difficulty and cost associated with this process when the consumer implemented the on-premise ERP systems, and this fear carries over to new Cloud adoptions.

- There is also the fear that software platform and application providers will introduce something in their periodic updates and releases to the Cloud solution that will change the data models or in some other manner deprecate or remove functionality that the consumer uses in business-critical functions. This shouldn't happen in any well-designed software, but the concern is a valid one.

- Current infrastructure platforms and solutions using x86, RISC, and Mainframe solutions may not all soon have Cloud solutions that can replace them. Resolving this barrier to Cloud entry is demand-driven, so the adoption and migration scenarios with the highest inhibition instances (opportunities) are likely to be first to receive attention.

- There may also be concerns around the quality of the vision of the provider, the ability of the provider's team to execute its vision, and the solvency, stability or financial ability of the company to continue to support the consumer's business without the disruptions of bankruptcies, etc.

Shapes In The Cloud

4

Shared Resources Serve Multiple Consumers

Virtualization, Multi-tenancy

Discussion Areas in This Chapter

- *Types of virtualization*

- *Challenges and opportunities for multi-tenancy*

Shared Resources Serve Multiple Consumers

Much of the value realized from the Cloud economic model is driven by the abstraction of computing resources. Abstraction in this case means that a single physical instance of a computing resource can be provisioned by a service provider to a number of independent using consumers, either in parallel for concurrent processing or in a serial fashion, where each user receives service at their expected level, unaffected by and unaware of the other users using the resource. The physical resources actually provisioned in the performance of service are abstracted to become transparent to the consumers.

A. Virtualization

An object in a mirror is not actually a physical object, but it is a reflection or image of the physical object. The image looks exactly like the actual object and looks to be in the same location. Virtual machine images or containers are similar to this. Virtualization in Cloud Computing might be defined as anything that serves to hide the physical characteristics of delivering computing resources from their other interacting systems, applications, and end users.

1. Types of virtualization

The following is a list of the most popular applications of resource virtualization.

- Platform virtualization:

  o Platform virtualization provides the simulation of machines for use by remote resources. The use of machine images or containers is often related to this. A virtual machine (VM) is a software implementation of a machine that executes programs like a physical machine.

o A system virtual machine provides a complete system platform supporting the execution of a complete operating system (OS).

o A process virtual machine runs a single program to support a single process.

- Resource virtualization has 2 basic cases:

o One case would be making a single physical resource appear to concurrently function as several resources, such as a web server, application server, and a storage solution.

o Another case would be making multiple physical resources such as servers or storage devices appear to function as a single logical resource, e.g. a single compute server or a single storage device.

Virtualization can be applied across all of the Cloud Computing framework, or just to one or more individual components of the framework. Here are some of the types of component-level virtualization that can occur, described only to introduce the scope of virtualization that is possible within Cloud Computing.

- Hardware virtualization encompasses various degrees of virtualization:

o Full virtualization requires the guest operating system's (OS) complete instruction set, input/output operations, interrupts, memory access, and any elements to be used by the applications software that is intended to run in the virtual machine.

- o Para-virtualization does not involve hardware environment simulation, but the guest OS/software is likely to require modifications that allow it to be run in an isolated domain.

- o Partial virtualization does not provide full environment simulation, so guest OS/ software may require some modifications to run as designed.

- o HW-assisted virtualization from the hardware itself may provide some support.

- Software virtualization encompasses:

  - o The virtual machine platform virtualization solution described above.

  - o Hosting 2 or more applications in a virtual environment separated from the physical OS instance.

  - o Hosting 2 or more virtualization environments on a single physical OS instance

- Memory virtualization consolidates the RAM resources distributed throughout a network into being provisional as a single memory pool.

- Network virtualization provides a network addressing space within or across networks.

- Storage virtualization completely abstracts the provisioning of logical storage units from physical storage devices.

- Database virtualization separates the database as a technical layer between the application layer and the storage layer of an application's technology solution stack.

- Data virtualization provides the presentation of data as a layer abstracted from the physical database systems, structures and storage components.

- Mobile virtualization provides system platform and process application program solutions for wireless devices.

- Desktop virtualization provides remote hosting, display and/or orchestration of a desktop computing environment.

2. Virtualization vendors

VMWare, IBM, Microsoft, Parallels, and NetApp are among the leaders of virtualization solutions.

B. Multi-tenant software

In virtualized environments, abstraction layers enable Cloud Computing infrastructure components to be run decoupled from the actual physical instances.

A basic multi-tenant software environment has a single physical instance of software that runs on the same operating system, on the same hardware, with the same data storage to concurrently serve multiple independent consumers. The architecture of a multi-tenant software application is designed to virtually partition its data and configuration, allowing each consumer to run on a customized virtual application instance.

One analogy for multi-tenancy in software might be where two or more tenants living in a common building share the stairs, halls, pool, laundry room, secured parking, and the building structure itself. They don't share their individual apartments, where they customize their own rented rooms to fit their needs and preferences.

1. One instance per consumer can be expensive

A multi-tenant architecture is quite different from a multi-instance architecture which concurrently supports multiple consumers through the allocation of separate software instances for each consumer. But this contrast serves to make the case for multi-tenancy: instances are expensive. Each instance requires its own maintenance, upgrade, and support for software, hardware, and infrastructure. The provider must sustain old versions, track instance specific patches, execute customer-specific regression testing, and provide complex migrations. Instance-level license costs for software can substantially increase this cost, but prudent use of open source software can partially offset this (e.g. LAMP web application framework software).

Multi-tenant providers reap the efficiencies in staff, resources, and infrastructure with a significant cost advantage over on-premise applications and single-tenant hosted solutions. The additional time, efforts and cost associated with designing a multi-tenant architecture is modest compared to the post-design benefits.

A multi-tenant software solution produces increased value for the service provider because it:

- Moves the model from a 1-to-1 solution, where one thing can only serve one consumer at any time, to a 1-to-many model, where any one thing can serve many consumers at any time.

- This significantly reduces the cost to support a large pool of demands, while dramatically increasing the potential income produced by driving the single resource instance closer to the goal of full capacity utilization, or minimum idle, non-revenue producing capacity.

A multi-tenant software solution produces increased value for the consumer because it:

- Means that the provider is constantly improving the software solution offering to reduce their related operating costs.

- Factors such as technology, market share, and growth opportunities are working within Cloud Computing to increase competition for the provider. Therefore, some or most of these cost reductions may be passed on to reduce the "pay-as-you-go" costs even further for the consumers.

2. Unique challenges of multi-tenant solutions

Software platforms must be designed and constructed to support the unique characteristics required for the development, implementation, and maintenance of multi-tenant applications. Some of the key aspects of this are:

- Consumer-specific configuration of access security and usage at the various levels of object and class types, data elements, screens, etc.

- Each tenant must be sufficiently extensible to allow the expansion of the data model and the creation of screens that will support the business of the consumer. Note that multi-tenant solutions use metadata extensively throughout the data model.

- Each tenant must be able to configure the process supported by the application to support their own end-to-end use cases and flows.

- Each tenant should be able to personalize the "skin" or look-and-feel of their application to support their company's policies, practices, and design guidelines.

3. Multi-tenant software vendors

Any platform that was developed exclusively to support Cloud SaaS applications, and any applications that were developed exclusively as SaaS Cloud offerings are very likely to be multi-tenant.

Some of the better known examples of multi-tenant software might be:

- Salesforce.com/AppExchange

- Google Apps

- SAP/Business Objects

- Microsoft Live Services/BPOS

- Sugar CRM

- Workday

- NetSuite ERP

5

Cloud Deployment Models

Characteristics, Impacts and Options

Discussion Areas in This Chapter

- *Private Clouds*

- *Public Clouds*

- *Hybrid Clouds*

- *Virtual Private and Community Clouds*

- *Super Clouds*

Cloud Deployment Models

Cloud Computing (refer to the earlier definition if needed) solutions can be deployed in several different ways. At one end of the spectrum, a single enterprise might completely own and exclusively operate all of its Cloud Computing solution resources for its own business use. At the other end, that enterprise might own no computing resources but instead rely entirely upon one or more service providers to provide all of its Cloud Computing solution resources through services the enterprise pays for as they are consumed. And then there is everything in between these two extreme cases.

The consumer's expectation in all of the deployment models is that the acquired Cloud services will be consumed in support of successful product and service offerings.

A. Private Cloud Definition

Figure 5-1 below represents a basic Private Cloud reference view.

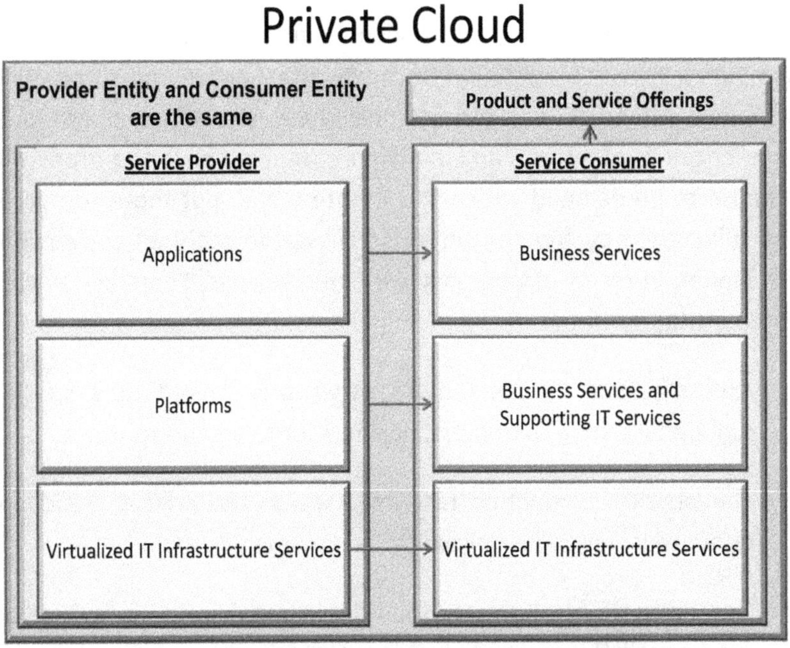

Private Cloud

Provider Entity and Consumer Entity are the same	Product and Service Offerings
Service Provider	**Service Consumer**
Applications	Business Services
Platforms	Business Services and Supporting IT Services
Virtualized IT Infrastructure Services	Virtualized IT Infrastructure Services

Figure 5-1: A basic Private Cloud relationship view.

An enterprise's first significant adoption of Cloud Computing will likely be to deploy a Private Cloud.

A **Private Cloud** is a Cloud Computing solution deployment that is centered on a single entity having exclusive availability and control of computing resources that it owns and controls. Activities and functions are provided as services within the organization's firewall over its intranet. Most enterprise network and business applications can be moved from their current on-premise solution basis to an on-demand private Cloud solution with relative ease. The applications need not be multi-tenant, but one would expect the new private Cloud's infrastructure resources would be fully virtualized to realize cost reductions and improve operational performance.

The values attained are a limited subset of the possible Cloud benefits, primarily associated with cost reductions from optimal infrastructure computing resource utilization (through virtualization and automation of owned resources) and process quality assurance from operational excellence. Applications and software can be converted from on-premise to on-demand within the Private Cloud, but these resources are still owned by the consumer, so these (dotted line) conversions are limited in terms of realizing real Cloud savings. Some key Private Cloud attributes are:

- Access to the Private Cloud's services is limited to a specific functional group, customers, partners, or to an enterprise.

- The physical computing resources within the Private Cloud are effectively and exclusively controlled by the Consumer.

- One example of a Private Cloud deployment would be where an Enterprise's IT operations build, operate, and own a Cloud for use only by consumers within the Enterprise.

There are several processes or computing activities (workloads) that are probable candidates for movement to a Private Cloud:

- Substantially-customized application-oriented workloads:

  o Enterprise resource planning (ERP) applications.

  o Industry-specific applications.

  o Process-specific applications.

- Database workloads:

  o Transactional databases.

  o Data mining, text mining or other analytics.

o   Data archiving and preservation.

o   Data warehouses/marts.

B. Public Cloud Definition

Figure 5-2 below represents a basic Public Cloud reference view.

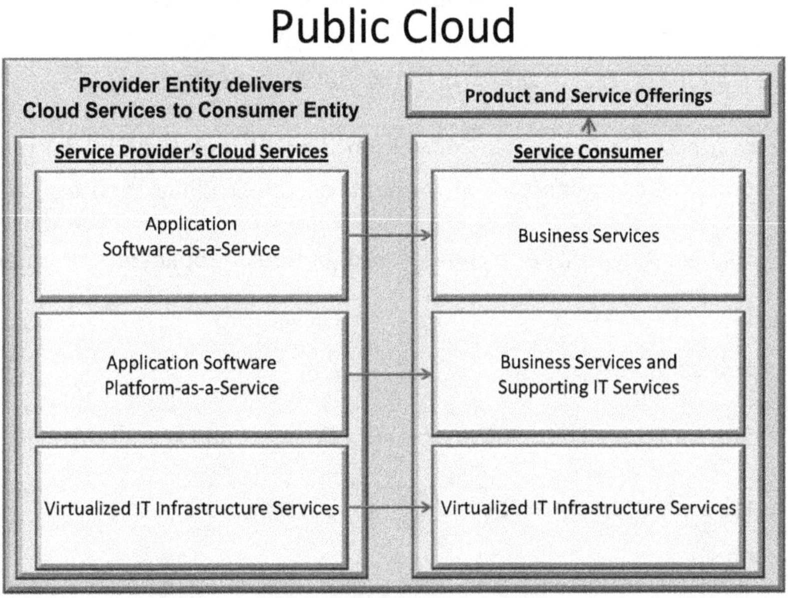

Public Cloud

Provider Entity delivers Cloud Services to Consumer Entity	Product and Service Offerings
Service Provider's Cloud Services	**Service Consumer**
Application Software-as-a-Service	Business Services
Application Software Platform-as-a-Service	Business Services and Supporting IT Services
Virtualized IT Infrastructure Services	Virtualized IT Infrastructure Services

Figure 5-2: A basic Public Cloud relationship view.

A **Public Cloud** is a Cloud Computing solution deployment where a set of IT activities and functions are dynamically provisioned through self-service over the Internet from external suppliers employing computing resources that are not owned by the consumer. The consumers are indifferent to the physical resources employed by the service provider for the Cloud services made available to them. The service provider abstracts its resources and supports concurrent demands from 2 or more authenticated public consumers.

Shapes In The Cloud

It is possible that an enterprise's first significant adoption of Cloud Computing will be to deploy services from a Public Cloud. Whether or not it is the first instance of an enterprise's Cloud Computing adoption, it is very likely that the enterprise will adopt one or more Public Clouds into its network of business services.

The service deployment might be for infrastructure services (such as Amazon EC2), for software application platform services (such as Amazon AWS, Force.com, Google AppEngine, etc.), or for software application services (such as HR, CRM, NetSuite, etc.).

There are several processes or computing activities (workloads) that are probable candidates for movement to a public Cloud service:

- Highly-standardized commoditized business application-oriented workloads

- Self-service apps

- Software application development platforms and ecosystems

- Infrastructure workloads:

   o Media streaming

   o Web conferencing

   o Software development, multi-platform support, testing, training, demonstration

   o WAN capacity and VOIP

   o Business continuity and recovery

   o Storage

C. Hybrid Cloud definition

Figure 5-3 below represents a basic Hybrid Cloud reference view.

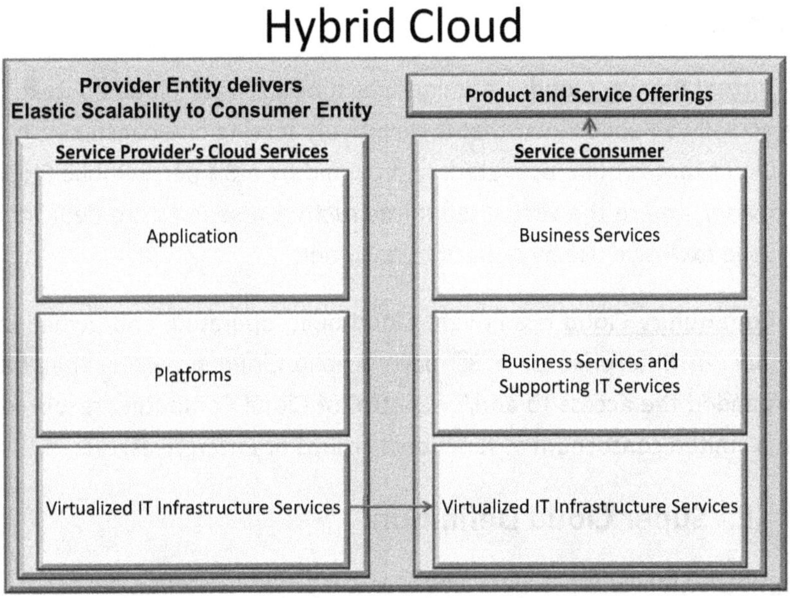

Hybrid Cloud

Provider Entity delivers Elastic Scalability to Consumer Entity	Product and Service Offerings
Service Provider's Cloud Services	**Service Consumer**
Application	Business Services
Platforms	Business Services and Supporting IT Services
Virtualized IT Infrastructure Services	Virtualized IT Infrastructure Services

Figure 5-3: A basic Hybrid Cloud relationship view.

A **Hybrid Cloud** is a Cloud Computing solution deployment model that provides one or more services resulting from the integration or conjunction of services delivered from a consumer's private infrastructure resources, and services delivered from an external service provider's public Cloud infrastructure resources. The individual supporting Clouds are separate entities integrated through technology enabling data and application portability. A spike or burst of compute or storage demands that are met through scalability from dynamic load-balancing of the Clouds is the primary example of a hybrid Cloud. A combination of public and private storage Clouds for archive and backup functions would be another example. Many infrastructure

Cloud Computing solutions are likely to be or will become hybrid models.

D. Virtual Private and Community Cloud Definitions

A **Virtual Private Cloud** is a unique form of a Private Cloud context. It is actually a Public Cloud that is exclusively serving one consumer. It is a Cloud that is built, operated, and owned by a 3rd party Public Cloud provider, where the virtualized infrastructure resources are dedicated for the exclusive use by a specific consumer.

A **Community Cloud** is a Private Cloud built, operated, and owned by either an Enterprise or a 3rd party Public Cloud provider, that has expanded the access to and/or control of Cloud Computing resources to a limited consortium of functional groups or Enterprises, etc.

E. Super Cloud Definition

Figure 5-4 below represents a basic Super Cloud reference view.

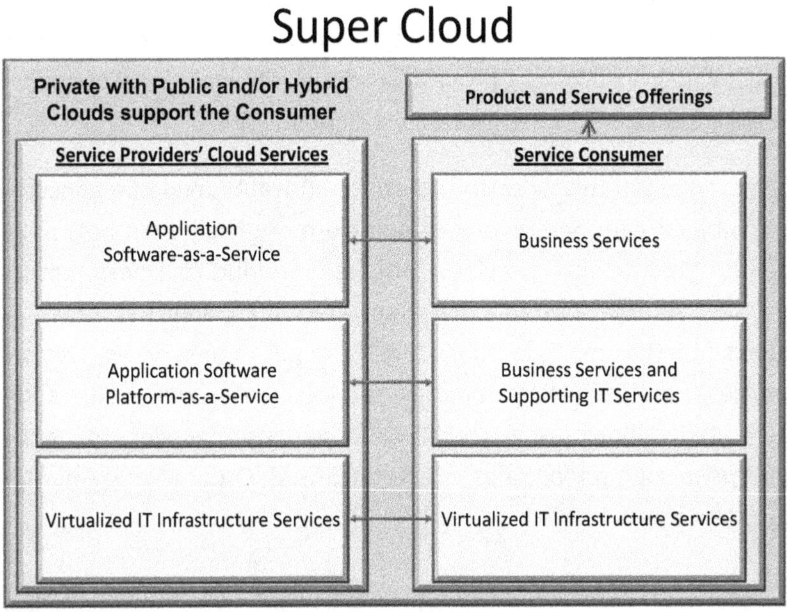

Super Cloud

Private with Public and/or Hybrid Clouds support the Consumer	Product and Service Offerings
Service Providers' Cloud Services	**Service Consumer**
Application Software-as-a-Service	Business Services
Application Software Platform-as-a-Service	Business Services and Supporting IT Services
Virtualized IT Infrastructure Services	Virtualized IT Infrastructure Services

Figure 5-4: A basic Super Cloud relationship view.

A **Super Cloud** is likely to become a dominant Cloud model for mature Cloud adoptees. There are several existing model names – Inter-cloud, Cloud of Clouds, Combined Cloud, Cloud Layer, Cloud Network, an interoperability Control Plane, etc. – that have been used to describe similar types of scope. The label "Super" is used here because it is simple and this description is intended to be sufficiently flexible to cover all of these other "instances". It isn't used here to start a new label standard, but as a catch-all model within this book for all other deployment scenarios not defined above. The reader is also apparently free to call this model whatever they would like to call it.

This model is a Cloud Computing solution deployment that joins together two or more Clouds for comprehensive end-to-end integrated business process support. This model does not include the hybrid Cloud model that uniquely supports infrastructures predominantly for elastic scalability reasons, but it does include all other multi-Cloud models. The integration of multiple Clouds' services to become one business process-centric IT service is one of the key challenges associated with the Super Cloud model.

One example of this would be where the Private Cloud of a consumer is combined with one or more (multiple more likely than not) public Clouds together to provide complete x-Cloud business process services. Another example might be where a Provider has effectively bound together multiple public Clouds' application services, platform services, and/or infrastructure services in some value-added integrated context as a specific offering encompassing the entire enterprise solution for Sales, Marketing, CRM, Order Management, e-Commerce, etc.

6

The Cloud Computing Overview and Framework

Layers and Components

Discussion Areas in This Chapter

- *The moving parts*

- *A Cloud Computing Framework*

The Cloud Computing Overview and Framework

There is a tremendous amount of interest in Cloud Computing from providers, consumers and users alike, and the momentum behind large-scale adoption of on-demand service-based computing solutions is expected to increase exponentially over the next 5-10 years.

A. The PC Analogy

Before beginning to drill down into some of the components that define Cloud Computing, an analogy would seem to be appropriate. Consider the parallels between Cloud Computing and the disruptive emergence of personal computing solutions in the 1980's.

IBM was a pioneer in bringing the original PC's to market. When they determined that their initial CPU and operating system components were inadequate for the potential of the PC as a new platform and form factor, they turned to Intel and Microsoft. The PC then became a "package" of components, opening the door for Compaq to take over market leadership. But over a period of several years after its introduction, various component vendors were able to drive standards and manufacturing equivalence to the point where all of the other PC manufacturers (the PC clones) took over dominant market share and market leadership in general. As product performance became fairly even among PC vendors, price became the primary differentiator, and value kept increasing while PC's became cheaper and cheaper.

The primary competition for PC's was Apple's personal computers, which always emphasized ease-of-use and the user experience combined with collaboratively-designed innovations to deliver a complete computing solution from scratch. Apple's specifications became their own product standards, whereby this one single vendor provided the fully integrated **solution** platform of CPU, operating system, video, storage, and memory components. For the first several years of competition with PC's, Apple's customers were willing to pay a substantial premium for computing that delivered real value because it actually worked well as an integrated solution. This premium would be drastically reduced only after the PC's major multi-vendor components became so standard, equal in value and price, that performance and ease-of-use differences became too small for the price premium originally demanded by Apple.

Cloud Computing, in some key ways, is in a very similar position. There are a lot of moving parts (components and layers) in Cloud Computing that must come together (integrated) in defining solutions that will be valuable to the target markets. The internet and web-based solutions provide a great springboard for the Cloud. And the first-to-market Cloud pioneers may be terrific innovators, providing great Cloud applications with consumers locked-in on the other layers and components. But it may also be that the "close followers" to the opportunity markets will pursue the correct partner alliances and effectively combine critical standards (e.g. security, virtualization, platforms) with improved quality, technologies, and performance to achieve broad adoption of mixed-vendor Cloud Computing solutions.

B. An Overview of the Basic Cloud Components

Cloud Computing requires a network, so it has many of the same moving parts as the on-premise web-based solutions that the consumer would otherwise buy, install, initialize, maintain, and manage for their desired web applications.

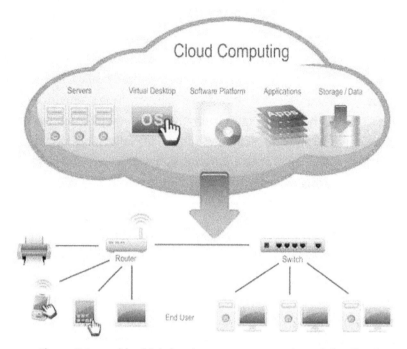

Figure 6-1 provides high-level component overview of the Cloud.

The Cloud includes several key points separate and apart from the typical web network considerations. It can be seen as the delivery by one or more service providers of a coherent, large-scale, remotely-accessible collection of abstracted compute, storage, networking, platform, and/or application resource services that have been requested by authorized service consumers and users. These services are deployed and allocated by the service providers using published API's over a network.

C. The Cloud Computing Framework

Beyond the basic component overview of the Cloud, there is a Cloud Computing framework that defines the layers of Cloud Computing, key cross-layer components, and the relationship between them.

Cloud Computing Overview and Framework

Refer to Figure 6-2 for a general representation of the framework. The framework clearly reflects its internet-based heritage. The framework component definitions are applicable to on-premise and web-based applications as well as Cloud on-demand applications. One thing that clearly sets the Cloud apart is that any one or more of these components in the Cloud framework may be delivered by a separate service provider as a collection of services to one or many consumers. There is an additional level of complexity borne by a Cloud Computing model over and above an on-premise solution model, in providing these resources as consumable services.

Cloud Computing Framework

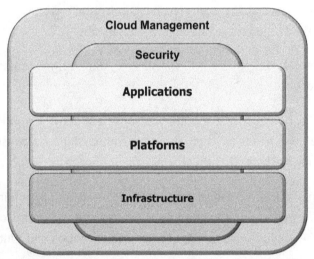

Figure 6-2 is a high-level view of the Cloud Computing Framework

In the practical application of any Cloud Computing solution, each framework layer:

- Abstracts the layer below it. A SaaS application uses the services that are provided to it by the application software platform that it runs on, with no regard for the physical workings of the platform.

The platform employs the infrastructure that it runs on without regard to the physical server, storage and network components and OS software instances allocated to its provision requests.

- Exposes interfaces that the layers above will build upon. The infrastructure layer provides an interface defined to meet every service request from the platform. The platform supports the application's interface requests for users, other systems, data, storage, etc.

- Provides no hard dependencies with the other layers, instead relying on the service architecture's solution of seeking, identifying, requesting, and applying services across the layers.

- Provides scalability and elasticity horizontally within its layer as required to meet the demand for its services.

The general definitions of the layers and components of the framework are:

- Applications provide a variety of business activity solutions that meet the needs of their users, encompassing any consumer or business activity that these users might engage in.

- Each application is developed, implemented, and maintained on a platform that enables every available feature and option delivered by the application at configuration, customization and runtime.

- The platform consumes the resources of the infrastructure to enable all of the applications' requests for services. The infrastructure layer encompasses the hardware and software components (servers, storage, network, operating systems and virtualization software) that are used to build, deploy, and deliver the Cloud Computing solutions architectural foundation.

- The management layer underpins everything as it provides performance assurance, integrity assurance, administration, operations management, and continuous improvement processes.

Although many discussions look at the Cloud as a pyramid, the framework structure works well as a combination of layers and components.

One analogy of highly-visible applications being supported by the less visible platform and infrastructure layers is the application of computing technology to allow modern aircraft to fly-by-wire. The pilot makes simple movements of user interfaces that are translated into thousands of movements in the airplane's control surfaces.

Some more down-to-earth Cloud perspectives think of the framework as an iceberg where only the application layer is visible to most people because it is above the visible "water line". Those people who can examine the iceberg below the water line might view the rest of the framework, and only they will understand that there is more to the framework beyond the application. Another Cloud perspective offers a similar analogy whereby the majestic swimming of a duck across a quiet pond is contrasted with the furious, mechanical paddling of the duck's legs and feet below the pond's surface.

These analogies may provide memorable visuals that there is a lot more than just the visible applications going on in successful Cloud Computing solutions, hence the need for a complete framework to drive the overall solution home. And that's it for water-related analogies.

7

General Cloud Computing Evaluation

Alignment, evaluation

Discussion Areas in This Chapter

- *Consumer's Cloud readiness*

- *Consumer's general evaluations*

General Cloud Computing Evaluation

Each layer and component in the Cloud Computing framework can be offered in the Cloud as a service. After all of the discussions above, it is appropriate here to provide some key insights and contents for making the evaluation of each of the elements contained within this framework.

We start first with a set of general evaluation parameters that can be applied by the consumer to most if not all Cloud Computing solution offerings, before we go into each individual layer and high-level component of the Cloud Computing framework.

A. Consumer Cloud Readiness Considerations

Every serious evaluation of Cloud offerings, especially those affecting the mission-critical operations of the enterprise, should begin with some general steps around a few high-level considerations. These are the first steps in determining if the solutions that are being considered can meet the performance expectations, if the solution's various vendors will effectively execute their roles in the consumer's success, and whether the consumer and providers are likely to be able to stand each other while making each other successful.

An evaluation is a key function of managing risk. A prospective consumer can substantially increase the probability that the choice they make will lead to optimal success if they put reasonable time in the preparation and performance of good evaluations. It is usually better to make an informed decision than an uninformed decision, while avoiding the negative impact of "analysis paralysis". These considerations basically encompass the definition of the consumer's expected Cloud success, the identification of the issues that must be addressed to achieve that success, and determining the vendor attributes that will make this success highly likely.

General Cloud Computing Evaluation

1. Defining / aligning the consumer's Cloud strategies

Whether the consumer is starting up its business and setting up its operations, or moving one or more of its resources from on-premise to on-demand support solutions, it should determine its core strategic vision for using the Cloud in its current and future business operations.

- What strategic business objectives are served by adopting Cloud Computing solutions?

- What business strategies need to be aligned for success? (Business charter, financial strategy, partner strategy, innovation strategy, technology strategy, market strategy, competencies strategy, product strategy, product line strategy, platform strategy, expansion and differentiation strategy, globalization strategy, etc.)

- Where will the consumer apply Cloud Computing support solutions? (Business processes, functions, computing resources, products, services, etc.)

- How will the consumer achieve its goals? What competencies will the consumer need to apply, acquire, develop, enhance, or drop? What types of significant additions or changes in resources will occur?

- What are the key factors that must occur or be present for the Cloud Computing vision to succeed? Why are these factors expected to favor the consumer's success over competitive environments?

2. Defining a Roadmap for realization of Cloud solutions

If the consumer is a startup, they may choose to get to full Cloud Computing adoption for all of their needs over a planned sequence of several steps or all at once. If the consumer has existing on-premise solutions supporting one or more of its mission-critical business processes, it is much more likely to get to its "end-state" of Cloud Computing through a number of calculated risk-mitigating steps.

In either case, the consumer will need to define a preliminary high-level roadmap and supporting plans that establish its milestone options or alternatives and their initial preferences. When performing its evaluation, selection, and initialization processes, it would be best if the consumer has more than a blank sheet of paper that is filled in by the various vendors. And they should have something more strategic than a comprehensive RFP.

3. Set minimum qualifications, priorities, and weighting

The consumer needs to determine a clear preliminary set of core requirements that must be met in the solution that they choose, other preliminary major requirements that are the highest priority beyond that minimum set, and an initial way for objective and subjective evaluation metrics to be applied in making a choice. These are neither extremely detailed nor complete, but they should ensure that the degree of risk associated with fatal surprises is drastically reduced if not eliminated.

B. Consumer Evaluation of the Provider

Some of the initial considerations focus on the service provider and are applicable to all layers of the Cloud Computing framework. There are several attributes of Cloud Computing resource or solution provision that the consumer should clarify with the provider, and the provider should be prepared to proactively address.

- Does the provider use its own Cloud Computing services to run its business processes? Does it use Cloud Computing services that are provided by other providers in its business operations?

- How stable is the service provider or its own service providers? Is it likely that they will be able to meet the consumer's desired levels of services? Why? If not, what are the possible remedies?

- Is the provider fully capable of supporting the consumer's Cloud Computing adoption Roadmap from initial release through one or more migrations to an end-state realization?

- Do the services and associated packages offered by the provider reflect innovation, market leadership or strong influence, technology leadership or strong influence, appropriate skills and competencies, clear competitive positioning and differentiation, etc.?

- How does the provider recognize new requirements? How agile is its solution enhancement and development methodology? What policies, tools and processes does it use to create, extend, drop, and deprecate, etc., its services?

1. Provider's management of partner providers

The service provider may be working with one or more partner service providers to deliver the Cloud Computing resources as demanded by the consumer. The service provider is typically responsible for all services provided under its service umbrella, assuring no constraints in service delivery and no additional liability to the consumer, regardless of whether the service is to be provided by provider or by one of its partners. The consumer should confirm this understanding and that the provider has the rights to deliver and bill for any of these 3rd party Cloud resources.

2. Support

The consumer still owns its data, the data is secure, and it is available at the expected service levels. Then the consumer must determine how the provider will support the consumer's operations in both the prevention of disruptive incidents, incident handling, and problem resolution.

a) What support does the service provider offer?

- **What are the types of support?**

    The consumer may require consulting to configure and customize the Cloud Computing solution to fit their needs. The provider may also apply event management to proactively seek improvements that prevent service disruption after the consumer is productive. Using the Cloud services as expected, the consumer may encounter disruptive incidents that will require the support of the service provider to remedy.

    The consumer must determine how to integrate their own IT service and support with that of the service provider. Understanding whether the provider has a call center, a helpdesk or a full service desk function to support the consumer's issues on a timely and effective basis is a good place to start. The consumer should also determine the channels of support, possibly encompassing agent calls by telephone, agent chat solutions, unique apps for mobile devices, a self-service portal, email, social media apps (Twitter, Facebook, YouTube, etc.). The consumer should confirm that there is a consistent, high-value consumer user experience across these channels, enabling case, incident and problem resolution through any and all of these channels regardless of where or how the incident was initiated by the consumer.

- **What are the hours of operations and SLA support commitments for the Provider?**

Support may be offered for a 10 hour period covering standard business hours, an extended period covering 18 hours, or 24 hours per day. The support may be for a 5-, 6-, or 7-day workweek, and it may or may not exclude holidays.

The consumer will need to map the hours of support from the provider to the SLA agreement and to the consumer's business operations around the globe.

- **What are the provider's service response and escalations policies and guidelines?**

An incident that disrupts the consumer's business operations cannot be resolved until the incident has been detected and the service provider has responded. The response encompasses a diagnosis period, a repair period, and a recovery period before reaching the service restore point.

The consumer and the provider should establish the committed response time and the expected maximum resolution time per incident, and the various methods that are available to the consumer to optimize this process. This should include any hierarchy of support escalations that may be required to ensure that the provider will not exceed the expected maximum time to restore the service to the agreed normal operation level.

- **How does the provider assure proper qualifications of the supporting staff?**

All of these agreements are negatively affected unless the provider has a sustainable commitment to staffing the support function with skilled and well-trained support professionals. The consumer should determine if the provider is successful in having the qualified staffing that is necessary to meet the SLA agreement criteria.

- **Is the provider committed to maximum uptime, minimum down, and customer satisfaction?**

  The consumer should be able to recognize and map the SLA service levels to the ways that the provider supports the consumer's business operations, and thereby determine the level of commitment that the service provider actually has to uptime and downtime requirements. If the consumer can see any inconsistencies between the SLA components and the actual support model provided, it should inform the provider of the concerns. The provider should be given the opportunity to remedy any shortcomings in an appropriate manner and in a reasonable time.

- **Does the provider have limitations on the number of support case submissions or hours?**

  The consumer should confirm that there are no limitations on the number of calls that it makes, the number of cases that it submits, or the number of solutions hours that are consumed to arrive at resolutions of its cases. The consumer should also understand what progression of steps are expected, e.g. that the user goes through the self-service portal first to get to known and common issues and their resolutions, etc.

3. Pricing, subscriptions, and metrics

Cloud Computing pricing should provide a broad spectrum of options that can be convenient and very flexible to support the consumer's unique financial needs. Metering of resource usage at all levels of granularity is available for the consumer, and is predominantly used by infrastructure providers. Period-based (e.g. monthly) subscriptions are typically associated with Cloud applications and platform offerings. The consumer should confirm how the provider will charge fees for partial periods (full period charge, ratably-charged for fraction of the period, no charge, etc.).

In any case, the consumer should work with the provider to establish a pricing package that will effectively and efficiently meet the consumer's usage, operational and financial management needs including potential reduced pricing or increased discounts for advance reservations or longer commitments. There should be no charges or fees from the provider to the consumer for maintenance, as these are not expected to have support for Cloud provider offerings.

4. Other contract issues

Basically, the consumer should seek a contract that preserves the various commitments from the provider, getting specific details from the provider that cover every possible contract option available for discussion.

There are some other considerations that are potentially worthy of discussions between the provider and consumer, and quite possibly inclusion in their contract.

a) What accounting, control, and financial issues may exist?

There are several unique aspects associated with Cloud Computing contracts. Any of the questions and issues that have been raised elsewhere in this book may be included in the contract between consumer and provider. But there are a few issues that are very specific to contracts that are addressed here.

- The provider should try to synchronize its recognition of revenues from the consumer with any expenses, including its costs from 3rd party partner providers, regardless of the timing of payments by the consumer or the provider's payments to its own partners.

- The consumer and the provider should each review their insurance policies for their business operations and make whatever changes and additions are necessary to reflect the on-demand model of their relationship.

- The consumer and provider should clearly establish any residual warranties, limitations, and exclusions expectations regarding service levels and performance, coverage period, and thresholds and caps of liabilities to either party.

- The consumer and provider should clearly define all causes for termination of the contract and the related service relationship. Conversely, the two parties should also define the methods and options available to either party for the cancellation of services, with or without cancellation fees or credits. A special condition that should be considered is whether a source code escrow should be considered for the event of the provider ceasing its business operations and ability to serve the consumer. In all cases, they should clarify needs and expectations for post-termination support e.g. transfer of data in an acceptable format for the consumer.

5. References

When the consumer is down to the last 3 competing Cloud Computing solutions (the finalists), it should be able to get the provider to provide access to 2-3 existing customers of each provider solution under consideration, with similar requirements to the consumer, whose success stories will hopefully increase the consumer's confidence in working with the provider's offering.

The consumer should inquire about any of the questions that are listed here, any of the subjects covered in this book, and any other points that are particularly critical to the consumer's success. The focus of these questions should deliver the needed insights into the actual behavior of the provider towards its consumers, especially behavior in the areas of greatest impact on the consumer's success.

Essentially, the consumer should seek to understand:

- How well the scope of the solution fits the expected depth and breadth of consumer expectations.

- The level of service provider integrity regarding expected execution and service levels.

- The reality of uptime and downtime performance.

- The issues that exist regarding interfaces, user productivity and consistent user experience.

- Financial surprise or major concerns regarding cost and value of the solutions.

Shapes In The Cloud

8

Evaluation Based On Cloud Computing Framework

General framework evaluations

Discussion Areas in This Chapter

- *Handling enhancements*

- *Integration options*

- *Usability*

Evaluation Based on Cloud Computing Framework

The best evaluation will be made contextually with the Cloud Computing framework. This permits the evaluation parameters associated with each layer's offering "as-a-service" to be considered within the context of the opportunities, limitations and challenges of the accompanying framework layers and components. For example, a platform-as-a-service (PaaS) offering may have limited infrastructure options and/or limited application development and runtime options that should be understood b y the prospective consumer.

Therefore, the sequence used below for the framework layers' evaluation discussions reflects a logical "reverse inheritance" bottoms-up approach of:

- Infrastructure services first, because these resources are directed by and support the platform(s), and evaluation parameters applied to them as Cloud offerings are likely to be key attributes applied to the platform.

- Next is the platform followed by the applications, because the platform enables the applications and the application evaluation parameters are likely to include some or all of the platform evaluation attributes.

- Security follows these layers because it must provide a holistic solution that encompasses all of these layers.

- All of these layers and the security components are supported by the management components of the framework.

Evaluation Based on Cloud Framework

A. General Cloud Computing Framework Evaluations

There are several evaluation attributes that can be generally applied across the framework.

1. Enhancements

The consumer should feel pretty good about the standard offering solutions. But the consumer will also need to understand how the provider expects to continue making improvements to its offerings and to ensure the ongoing use of the latest technology in its offerings.

Hardware resources may be predominantly isolated within the infrastructure layer, while software is widely used throughout all layers and components of the framework. That is why we enhancements are here, as general requirements applicable to the entire framework, rather than distributed into the respective layers and components below.

a) What is the frequency of updates made by the provider to its Cloud offering?

One of the great values of Cloud Computing is its agility. The key goal here is shorter enhancement and update cycles that transparently roll-out new technologies, improved performance, and/or new valuable features several times per year, while requiring little or no effort from the consumer and producing no disruption in service for the consumer.

Hardware updates can be made by the provider at almost any time, as long as the SLA's terms are being met.

Software updates typically follow a cycle that reflects the development methodology used by the provider. Most pure Cloud Computing software providers utilize an agile or lean method, such as Scrum, XP (Extreme Programming), etc. They will follow a cycle of iterations for development, where these iterations could be weekly, bi-weekly, monthly, and quarterly, etc. Some of these providers will hold these iterations back from enhancing the production code line, making less frequent releases that combine 2 or more iterations' development.

Other software vendors have come from an on-premise portfolio, where they are more likely to use a Waterfall methodology with fewer releases, to now also offer Cloud Computing solutions that use Scrum or some hybrid of Scrum and Waterfall.

The most common frequency for Cloud software updates is probably monthly, with the longest Scrum cycles probably 3-4 months, and the longest release cycles probably 1 year.

b) Are these updates automatic or can a consumer opt-in or opt-out for a period of time?

A consumer may want to be able to have some control over if, how and when they adopt the latest enhancements to the services they are consuming. With hardware, this is very unlikely unless the deployment is a private or hybrid Cloud.

With software, there is a wide spectrum of possibilities here that the consumer should get cleared up. On-premise software typically provides a great deal of latitude for the software owner to determine if and when they will implement upgrades. Cloud Computing software is usually far more restrictive, because great Cloud economy can come from supporting only one code line.

Evaluation Based on Cloud Framework

A Cloud software provider may provide the consumer the opportunity to be an early tester and adopter of its enhancement updates; or the opportunity to opt-out for a one-month delay in adopting the updated version; or no opportunities for anything other than taking the updated software whenever the code line updates occur. The consumer needs to understand what their options, if any, are, and whether these options are acceptable or not before committing to a specific software provider.

c) **How does the provider ensure that enhancements in new updates or releases will not negatively impact the consumer's existing business processes?**

Existing service level commitments should not be affected by any actions taken by any provider. With hardware resources, the service level commitment has been well-understood for quite a long time, and any changes to infrastructure will not be taken in ways that substantially introduce risk to the business consumers.

Well-designed software solutions should also respect service level commitments, current possible configurations, and the hardened exit points for consumer customization of their platform or application software. The provider should never deprecate or remove functionality, data model attributes, process model options, etc.

d) **Is a published product roadmap available to the consumer, perhaps under NDA?**

It is helpful for the consumer to be aware of what enhancements are coming and when they are coming into the services they are consuming. For hardware, this information helps the IT functions of the consumer understand what technologies and economies are being used by the provider, and how the provider continues its commitment to improve the quality and/or cost aspects of its offered services.

For software providers, a roadmap is much more likely. The consumer should be able, especially if under NDA, to get clear expectations set regarding what new features will available in the standard service and when, because these new features will enable the consumer's own expansion in its business model, its own quality improvements, and its own cost improvements. It gives the consumer a sense of the provider's ability and reliability in making continuous service improvements. So the consumer should definitely seek clarity on what the provider will give to the consumer in this context.

e) How will the consumer participate in the requirements gathering, analyses, prioritization, and implementation review activities of the provider?

The consumer should determine how they will be able to influence the provider's roadmap in its development of new features that are critical to the success of the consumer. There are a variety of possible channels for this influence, including submission of bug reports and enhancement requests; provider and consumer visits to each other's business sites; focus groups, advisory boards and councils, where groups of consumers selected by the provider review and advise on features and priorities; and bleeding edge alpha and beta testers, early adopters, etc.

The consumer should determine what channels for influence are available, and which of these they will use.

2. Integration

Integration is predominantly accomplished through software that affects all layers and components of the framework. Hence, it will be discussed here rather than distributed to each layer and component section below.

Whether the consumer is a startup initializing its entire operations on Cloud Computing services, a SMB moving to the Cloud in a "big bang" implementation, or a mature larger business moving in a carefully orchestrated multi-step progression to the Cloud, integration will be a very critical aspect of its Cloud Computing success. In fact, quite often, it is the "cost" of integrating in the Cloud that may be the most unwelcome surprise in Cloud projects, and the greatest culprit leading to failure in some form relative to expectations.

a) Does the provider's offering deliver and support published APIs for exposing its processes and model to other systems?

Software APIs (Application Program Interfaces) provide the bridge between the Cloud Computing solution and the consumer's other platforms and applications that it uses to run its business operations. Integration of business applications has been one of the great compelling benefits produced by predominantly on-premise ERP software suites over the past 15-plus years. So it is difficult to reasonably expect that a consumer would actually lose effectiveness in its end-to-end cross-application or cross-platform business operations just to get the other advantages of the Cloud.

The consumer should absolutely determine if the Cloud Computing software provider has one or more API's for integration, and what each API provides relative to the consumer's specific integration requirements, prior to committing to a specific provider.

b) Are there existing libraries for standard integration to specific external on-demand and on-premise solutions?

If the provider has created a library of standard services that the consumer can use to construct its desired business processes, and the services needed by the consumer already exist, this will make the time-to-production much faster for the consumer. The consumer should get a complete definition of all available libraries, services, and other loosely orchestrated components that the provider has standardized for the purpose of improving the implementation by the consumer.

c) What are the steps that a consumer would have to go through to actually implement an integration solution?

The provider may have the API and the libraries, but its integration tools or support for the desired development environment may make the integration process itself easy or difficult. The consumer should determine the actual steps for each of the integrations that it expects to make, and be informed before committing to a specific provider's solution.

3. Usability

Usability is directly associated with software in most Cloud solution cases, so we will be discussing it here rather distributing the discussion in the layers and components sections below.

The user's experience in performing their assigned functions is also an important area for evaluation. If the Cloud Computing offering requires user interfaces, they should be designed with the user persona and optimized user interface in mind. Productivity, the ease in attaining results, and user appreciation are critical to the success of the project and the business adoption of the Cloud Computing solution. User experience impacts the consumer's cost of doing business and it can make or break the consumer's agility to adopt enhancement and new business solutions, which is critical to the success of both consumer and provider.

The goal should be that the provider's offering reflects a user interface design that is most effective for the intended use, consistent within the Cloud offering, and consistent with generally accepted best-design practices in other well-liked web solutions.

a) **Does the provider deliver free trial access and support for evaluation of usability by key consumer users?**

The best known way to evaluate the effectiveness of graphic user interfaces is to have a combination of:

- Analysis and research specifically on the design provider's interfaces across its software that is being considered by the consumer. This analysis compares the various parameters of well-designed interfaces, the user experience design and performance policies of the consumer's enterprise, to the interfaces of the software.

- Hands-on use of the software in support of the consumer's business processes by the consumer's subject matter experts, preferably the representatives of the same people whose productivity is directly affected by the software under consideration. Trial access on a formalized basis should benefit both the consumer and the provider on getting down to any critical GUI issues before commitment to the relationship.

b) Does the provider's offering require extensive training for the expected user persona to use?

The user experience research and analysis available to the consumer is the best source of usability insight for the consumer. Another key indicating source is the amount of training for each affected user that the provider recommends for success in using its offerings. If the training is extensive, and the processes are considered relatively straightforward, this is likely to be an indicator that the software is probably not very easy-to-use. Access to other consumer companies' users for their insights would also be helpful.

This issue can seriously affect the consumer's TCO (total cost of ownership). Software solutions that are difficult for users to implement, adopt, maintain, and adapt to new updates will have a higher cost to the consumer over time.

c) Does the provider use interface technologies for easy-to-use cross-platform support?

Cloud Computing applications can be presented to users through computers and mobile devices. If the consumer is interested in supporting its interfaces for multiple access types of devices, the consumer should confirm the scope of the standard support for multiple access platforms, and the extent that the technologies employed by the provider minimize the burden on the provider. If the interfaces have used Flash, Silverlight, or other RAIA-based (rich anywhere interface application) solutions, single lines of code for the interfaces should serve to reduce the differences between the consumer access channels, reduce the costs to the provider in developing and maintaining their interfaces, and again increase the productivity of the consumer's users.

9

The Infrastructure Layer

Components, Options, Services

Discussion Areas in This Chapter

- *Servers, storage, network, desktop*

- *IaaS*

- *Candidates*

The Infrastructure Layer

The first area of focus within the framework is the infrastructure layer, because it is the layer that the rest of the framework runs on to build and deploy operational solutions.

A data center is a good representation of the physical machines, operating systems, network systems, and storage systems that are defined as infrastructure elements. The true costs of operating an owned data center should also consider the expected or experienced costs associated with the space or real estate required, power and cooling, backup generators, procurement, installation and certification of servers, maintenance, decommissioning, and security. The physical infrastructure resources may be consumed directly or as services offered within a Cloud context. Note that many of these components may have their own individual SLA parameters, but the Cloud Computing infrastructure's best SLA attributes may be those around end-to-end global enterprise-class availability, performance, and security.

A. Servers

The client–server model is critical to Cloud Computing. Servers provide the computing power that moves data across a network to enable applications, operating systems, file transfer, email, printing and communications among clients.

Each unique type of server is defined by its configuration and software, managed by a platform that pairs specified hardware and software to deliver data transfer for reliably scaled computing power across incoming client requests. Defined communication roles improve the server's ability to secure data for its intended recipients.

The Infrastructure Layer

Cloud Computing requires a great deal of flexibility and agility to deploy computing resources on demand. The right types of servers and configuration permit an agile structure to quickly implement new ideas, features or services, and an effective way to get more use out of the server investment. Distributed computing allocates processing power and data transfer in the most effective way.

1. General server considerations

There are several general data-center and server-related considerations that are worth mentioning within the Cloud architecture context.

- Energy-efficient power management begins with data centers placed in efficient buildings that are equipped with climate control systems and green servers which together produce "zero- or low-footprint" solutions, and can deliver up to 50% reductions in energy costs. When you add the related efficiencies from much greater utilization rates at Cloud data centers, you have tremendous economic, efficiency and sustainability improvements.

- The modularity that Blade servers provide contributes to ease of upgrades required by Cloud Computing solutions. And the smaller footprint improves the productivity of the data center.

- Server-farms with limited budgets like to invest in new server capacity as demand grows. Rack-mounted mobile servers that are built into cabinet containers will allow an ease of upgrades as well.

2. Server configuration options

Using the right server configuration to support a Cloud Computing solution can produce substantial savings.

- Shared server configurations are the most economical form of supporting multiple varieties of users, as user access is partitioned but memory and transfer resources are shared.

- Virtual server configurations provide segmented servers provide users with full root access and computing, effective for development environments without dedicated servers.

- Dedicated server configurations deliver more computing, as they provide full computing from a given machine with improved engineering monitoring and support.

- Cluster server configurations provide a basic distributed environment where a set of machines are dedicated to distribute content or requests.

- Grid server configurations extend on the cluster by combining multiple servers to act in unison with each other, making this attractive for business applications that are mission critical.

This configuration provides the most efficient use of dedicated hardware on a network, and it is the optimal server configuration for the Cloud. It provides distributed computing with load balancing, scalability and elasticity across a network of multiple servers supporting specific application roles (compute, database, storage, transfer, etc.). Over-the-top vendors such as Google and Amazon offer their server Clouds under this configuration.

3. Server application categories

Any given hardware setup can often power multiple applications on the same machines, but the application of servers to specific application roles has enabled specialization and improved uptimes, efficiency and reliability of networks. Data centers are often configured around specific purposes for greater efficiency.

- Web servers are optimized for data transfer speed (VOIP and streaming media), security with extensive firewalls and incoming request filters. They host FTP files and web sites, and deliver images, text, rich media and data over the Internet.

- Application servers are involved in complex web processes related to specific software, such as business applications, games, output from scripts and program, serving API (application protocol interface) data. They are usually optimized for load balancing and distributing data efficiently.

- Media servers provide an efficient way to transfer high bandwidth media files online, with substantial amounts of RAM and multi-core CPUs to maximize data transfer rates.

- Exchange servers typically use SMTP to provide an outbound client and inbound retrieval server for electronic messages, applying set rules, filters, checks and exclusion lists to protect and successfully deliver messages

- Other types of specialized or dedicated servers include ftp servers, database servers, storage servers, name servers, file servers, print servers, and terminal servers. It should also be noted that many web services are types of servers.

4. Top server vendors

Server revenue in 2010 appears to be approaching a 15% growth rate over 2009, primarily driven by the increasing adoption of x86-based architecture systems over the RISC-based systems.

- HP leads the market with around 30% share, followed by Dell, IBM, Fujitsu, and Oracle/Sun.

- 3rd quarter 2010 shipments of Rack servers had 24% growth with a 31% growth in revenue

- 3rd quarter 2010 shipments of Blade servers had 7% growth with a 26% growth in revenue.

B. Storage

Data storage occurs in many different ways using various persistence technologies.

- Memory data stores are called either primary storage or the first level of computing storage, and it is directly accessible by the CPU.

- Hard disk drives are the dominant type of secondary storage, allowing the CPU to indirectly access the storage by using input/output channels to temporarily utilize primary storage as an intermediary.

- DAS is direct-attached mass storage that can be accessed through an indirect method.

- Tertiary storage involves the use of removable mass storage that is mounted and dismounted from a storage device according to the instructions of a CPU over a network.

- Offline storage is mass storage on a media or device that is not under the direct control of a CPU.

- NAS is network-attached storage that is attached to and managed by a computer. The storage can be accessed directly at the file level by another computer using a network connection (over a LAN local area network, a WAN wide area network, or the Internet). NAS manages the presentation of files systems to its client computers.

- SAN is a storage area network that provides storage capacity to other computers. The storage can be accessed only at a raw block level, requiring the attaching computer to manage the file systems and data acquired from the SAN network.

1. Major Cloud storage categories

There are three fundamental types of Cloud storage services. Cloud storage services are abstracted from the physical storage resources, and exposed through well-defined storage interfaces. The service level agreements should establish data durability and availability levels.

- Durability defines the extent that any stored object will remain intact and accessible. Amazon's default of 99.999999999% guarantee means out of 10,000 objects stored the storage provider may lose one of them every 10 million years or so. Amazon's reduced reliability storage (RRS) rating of 99.99% guarantee means out of 10,000 objects stored the storage provider may lose one of them every year.

- A storage availability rating of 99.99% or better should be the goal, but the consumer should recognize that there are a number of points of failure for an attaching computer's storage availability.

a) Basic services

The most basic of storage services is plain data stored in a remote storage Cloud. Amazon S3 (Simple Storage Service) is an example of this coupled with its Cloudfront service to reduce network latency. One great example of this is the storage provided to NASDAQ exchange. Storage service consumers will use the provider's API to connect to storage SOAP and REST web protocols or virtual file system gateways to storage networks.

b) Backup services

A number of companies offer dedicated Cloud backup services that include scheduling, reporting and file recovery. For example, Amazon offers its RDS (Relational Database Service) for MySQL backup.

For example, Amazon offers its RDS (Relational Database Service) for MySQL backup. Other offerings include Carbonite's Online Backup, EMC's Mozy, and Symantec's Online Backup. Some backup service providers are actually packaging the Amazon or other vendors' backup services.

c) Archive services

Some vendors offer a Cloud storage service tailored to the needs of data archiving, including search, guaranteed immutability, data lifecycle management, deduplication, setting read-only or write once read many (WORM) status, e-discovery, etc. A great example of the application of an archiving service is one that comes up in many conversations and web searches on the storage topic, where the New York Times uses Cloud storage to archive and render its entire library of articles.

2. Top storage vendors

According to the latest results available in 2010, the disk storage market has been fairly recession-resistant with growth nearing 20%. Its leaders for the 3rd quarter of 2010 include EMC as the top vendor with around 27% market share, followed by IBM at around 13%, NetApp and HP tied at 11%, and Dell at 9%. NetApp had the greatest growth rate at over 50%. HP's acquisition of 3PAR and EMC's acquisition of Isilon should be reflected in the next round of results announcements.

The Infrastructure Layer

Some of the best candidates for individual enterprises building Cloud storage on a massive scale behind their own firewalls might be NetApp / Bycast's StorageGrid, Cleversafe, and EMC Atmos, with dispersed physical storage using object-based addressing versus using file systems.

The following is a list of some of the most popular storage vendors revealed through searches on the web:

- ZumoDrive is a Cloud-based file synchronization and storage service. The service enables users to store and sync files online and between computers using their Hybrid Cloud storage solution.

- Wuala is a free secure online storage which allows its users to securely store, backup, and access files from anywhere and to share files easily

- Egnyte delivers an all-in-one file server solution that combines unlimited, on-demand file storage and workgroup collaboration capabilities with an automated, online backup.

- Jungle Disk is an online backup tool that stores its data in Amazon S3 or Rackspace Cloud Files.

- Box.net is a Cloud Content Management (CCM) platform.

- HostedFTP currently uses Amazon S3 to store customer files and Amazon EC2 to host their website and database, and they use Amazon EC2's EBS (Elastic Block Store) to manage codebase. This system allows them to quickly deploy new versions of their software with zero downtime and without having to create new EC2 machine images.

C. Network

The network is critical to any form of Cloud Computing solutions. Routers, switches, and unifying communication systems' platforms are needed to deliver scalable, on-demand computer cycles, and the concept of service provision is intrinsic to a computing network. Cloud services must be provided over a network to a consumer, and these services cannot be economically feasible without some scale, which in itself means that there must be a network that can enable that scale. Server farms would not be financially plausible without networks. Distributed storage is by definition attached to a network.

Within the Cloud Computing context, there may be platforms in place that manage each of these traditional services. But the greatest Cloud difference comes from the realization that all resources -- infrastructure, platform, and application -- are being made available over a network. Interconnections, communications, integration, security, etc., are obvious Cloud network services. In a more expansive context, the Cloud network services focus on managing the entire network's resources from a holistic perspective, end-to-end rather than point-by-point, while still enabling operational and process improvement insights into the performance of the Cloud's components on a particular business service.

1. Network definitions

Networking is the connection to and from shared resources, systems and services from anywhere that a user can gain authenticated access. Network services are provided to a customer within a business to deliver interconnectivity of the consumer's unified computing systems, resources and other services. These network services must provide the agreed performance levels on a consistent and reliable basis, with appropriate security and on-demand scalability to support the customer's distributed business computing processes.

The Infrastructure Layer

a) Communication protocols

The OSI (Open System Interconnection) model defines a network framework for communication protocols across 7 layers. Communication control is passed from one layer to the next, starting at one network station's application layer, moving down to the station's bottom layer and over the network to the bottom layer of another network station, and up the layer hierarchy of that station.

As stated earlier, Cloud Computing is based on internet technologies. The internet network protocol is commonly referred to as TCP/IP (Transmission Control Protocol / Internet Protocol). Although it wasn't designed with security in mind, it provides the internet's communication protocols across 4 layers.

- The **Internet Interface** layer provides the direct communication to and from the actual network hardware (e.g. the Ethernet card).

- The **Internet** network layer decides where and how the data should be sent to get to its destination.

- The **Transport** layer provides and handles the data flows for the application layer, with guarantees of reliability.

- The **Application** layer is where users interact with the network.

Figure 9-1 provides a mapping of the OSI framework model's 7 layers to the 4 layers of the Internet protocols of TCP/IP.

Protocol Map TCP/IP to OSI

TCP/IP Transmission Control/ Internet	OSI Open System Interconnection
	Application
Application	Presentation
	Session
Transport	Transport
Internet	Network
Internet Interface	Data Link
	Physical

Figure 9-1 provides a mapping of TCP/IP and OSI layers.

b) Basic network computing components

The basic activity in initiating a service begins with a human or system "user" triggering a web service request which "calls" the Cloud. The Cloud then goes through several steps to service that request:

- It accepts the request.

- It confirms that the requestor has permission to make the request.

- It validates the request against rules and controls (account limits, etc.).

- It locates and identifies physical resources that are qualified to meet the request and free to meet the request.

- It binds or provisions the resources to the requestor's account.

- It initializes the resources to meet the request.

- It returns resource identifiers to the requestor (application) to enable it to meet the request.

- The application has exclusive access to the resources for as much time as needed.

- When the application has completed its processes, it returns (releases) the resources back to the Cloud service provider.

- The Cloud prepares the released resources for reuse by reformatting, erasing, or rebooting as needed, and then sets the resource status as "free".

From the perspective of a Cloud network, all of these steps require services that enable the computing resources to meet the request. The service subscriber (user) uses the application interface to make the request. Then the virtualized Cloud infrastructure responds with a virtual machine image that contains (at least) the called application and an appropriate operating system (usually a guest version of the OS).

The VMI moves through the **network** using virtual components (e.g. a VSwitch or virtual switch) to get to the server hardware computing resources required for the process, while authenticating access permission, identifying the desired storage and database resources, determining the computing requirements (e.g. massively parallel grid computing), finding network peering for least-cost internet connectivity, and on to access the internet network and other Clouds through the enterprise's internet service provider.

Network service providers use a combination of hardware and software services to deliver performing, secure, reliable, and scalable networks. A list of some of the software services provided would include:

- A Cloud Infrastructure Management Platform, usually operating on a network infrastructure resource configuration separate from the business operations network resources.

- Resource virtualization software from VMWare, XEN, etc.

- VSwitch software from Cisco, etc.

- Storage virtualization and allocation, from NetApps, EMC, etc.

- Unified computing systems solutions from Cisco, Oracle, HP, other server vendors, etc.

- Authentication and access providers.

- Application communication environment support

- Wide area application services and application velocity solutions to accelerate remote application processing.

- Intrusion detection, distributed denial of service, secure socket layer support.

- Site selection, load-balancing and cluster management.

c) Traditional LAN or WAN enterprise network services

Traditional network services ensure security and user friendly operations over a network, helping the network run smoothly and efficiently. Some of the most common services are:

- DNS (Domain Name System) to give intelligible name labels to network resource addresses.

- DHCP (Dynamic Host Configuration Protocol) eases the network administrative burden by automating the IP assignment of nodes on the network.

- Authentication servers allow every user to have their own account with a registered user name and password, making users accountable for all of the logged activity that they do on the network.

- E-mail, printing and network file sharing services are network services that allow users to use a corporate mail service, access any printer connected to the network, access files on the server or other connected nodes, and streamline data transfer within the network.

- Directory services require users to have permissions to access the shared resources through easy-to-configure security and access rights.

2. Network services vendors

Some of the leading providers of network services are listed below in random order:

- HP Network Services

- Cisco Network Services

- Level 3 Computing Services

- Amazon Network Services

- IBM: Cloudburst, Network Services

- AT&T Network Services

D. Virtual desktop infrastructure

Cloud Computing solutions that deliver virtual desktops provide several benefits to the business IT function and its enterprise. Moving to virtual desktops provides a form of client virtualization to improve local security, increase file storage economies of scale, and apply desired protocols uniformly and consistently across an organization.

Some of the most typical examples include the following:

- Providing cross-platform and centralized virtual desktop infrastructure (VDI) solutions delivering a hosted operating system in a virtual image that permits users to run guest applications they require to perform their business duties effectively.

- The development and maintenance of a centralized role-based image for each distributed desktop software configuration that is required by the users within an enterprise. This will provide:

  o Reduced cost to install, maintain, upgrade, transition, and transform platforms and applications, because only types are maintained instead of individual instances.

  o Faster time to onboard and bring to productive status new applications and new users.

- Support client access to application streaming from a host to the guest computer.

- Quest Software is an interesting vendor in this space.

E. IaaS: Infrastructure-as-a-Service

There are a variety of estimates regarding IaaS revenues and growth. Projections for 2013 IaaS service provider's revenues range from $10 to $12 B., 20 to 25% of the global Cloud Computing revenues.

IaaS is a computing solution whereby the physical resources of the Cloud Computing service provider's infrastructure are abstracted through <u>virtualization</u> to allow automated resource provisioning, and/or support platform and application partitioning through <u>multi-tenant</u> architecture. The service consumer configures the infrastructure's virtual machines to meet its business needs, and deploys the desired business applications and workloads on them. The virtual resources are delivered by the provider to the consumer as standardized services over a network through APIs that manage the resource abstractions.

The consumer's IaaS adoption and migration is driven by applications, likely initiated by a consumer's business management team recognizing that moving one or more applications to an IaaS arrangement will produce worthy expected <u>application-driven benefits</u>. These benefits could be as diverse as cost reductions and increased agility from application templates instead of multiple instances, and increased revenue through shorter times to productive sales, marketing and business development applications.

F. Primary business candidates for IaaS

Some types of applications are better candidates than others for earlier Cloud Computing adoption. They reflect the stress placed on current business processes that are "spikey" or "bursty", placing extraordinarily high peak demands or infrequent unpredictable demands; or there are demands that require agility and elasticity to quickly roll up and roll down resource requirements. The Cloud demand patterns of these candidate applications vary across compute intensity, storage density, and communication capacity.

- Application development requires substantial investments in infrastructure to support all of the various possible platforms, configurations and test cases that must be validated. An <u>extensive discussion</u> of these types of demands is provided in a separate section.

- The server, storage and network demands associated with file management and print management.

- Process management demands can be fairly extensive if done correctly. Merging, enhancing or re-engineering processes and their associated systems require extensive testing and staging. Add to this spectrum of resource demands that support Business Continuity and Disaster Recovery, and there are a number of very attractive drives for IaaS solutions here.

- Computer design, modeling, scientific projects, and engineering optimization initiatives can prodigiously consume infrastructure resources due to their very research-driven nature.

- Whole systems for decision support, including frequent dashboard refreshes, reporting and analytics, and the associated data warehousing demands, are great candidates for IaaS.

- The production systems that support the core business processing.

o Some systems, like ERP and specific applications like CRM, HR, finance, control, accounting, etc., can have fairly predictable high-volume demands that can be placed on IaaS.

o Batch processes are unique examples of business activities that can be easily addressed by IaaS resources.

o Transaction processing systems (OLTP) and web support systems are examples where the consumer demands can become highly variable and very unpredictable, and are therefore great candidates for IaaS.

o Email, web conferencing, and other collaboration activities are easily separable and moved to IaaS support.

o Virtual desktop media management and storage support for audio / video streaming are other growing areas of interest for IaaS.

G. Key IaaS Evaluation Attributes

The evaluation of IaaS providers' offerings emphasize agility, service levels, automation, deployment model options, scalability and elasticity, virtualization, multi-tenancy, etc.

1. Infrastructure delivery

a) Does the provider support all deployment models?

The consumer should determine if the provider has any preferences, limitations, or success ranking associated with the various possible Cloud deployment models. The consumer may want to start with a private Cloud deployment for a limited number of applications, but there are too many public Cloud solutions that can handle standard business functions to believe a public Cloud application is not in their future. In fact, it is much more likely that there will be a multi-Cloud landscape with hybrid Cloud solutions that have greater deployment density for any consumer.

b) Does the provider offer any forms of turnkey solutions?

It is a good thing when a provider can offer or support the creation of libraries of predefined design solution components with broad reusability, to produce faster and easier deployment of the consumer's desired applications with quality and integrity. Spinning up new deployments rapidly can mean the competitive difference between growth and decline in business for the consumer.

2. Integration capabilities

a) Does the provider support integration with other Cloud vendors?

It is possible, given sufficient time and money, that everything a consumer needs can ultimately be provided by a single provider, and that the level of integration will be terrific. But in this world that demands pragmatic agility, loves portability, and loathes vendor lock-in, it is extremely unlikely that such a scenario would be in the best interests of the consumer.

The consumer should determine if there are any integration limitations with other Cloud vendors' solutions.

b) What are the integration solutions for end-to-end cross-platform processes?

Not everyone and everything will be residing in the Cloud. The consumer is probably going to have processes whose end-to-end mission-critical processes encompass multiple on-demand and on-premise applications across Clouds, across data centers, across partners, etc. The consumer must confirm that the scenarios it might expect, from least to most likely, are or are not supported by the provider as a critical aspect of its risk management.

3. Availability, Reliability, Maintainability and Metering

So let's assume that it is still the consumer's data and it is secure. One needs to also determine if the operations of the provider are sufficient to deliver the availability that the consumer expects for its Cloud operations to continue to succeed. I have a colleague who uses a tag as part of his professional self-introduction that goes something like this: "I take in great volumes of demand requirements and spit out nines". This seems applicable here.

a) **What are the provider's guarantees and supporting published statistics regarding uptime availability?**

The consumer and provider will define the expected service levels. The consumer should be completely aware of the expected uptime availability guaranteed by the provider, and the consumer should seek the provider's statistics that will support the provider's guarantee and bolster the confidence of the consumer.

For software platforms and applications the minimum uptime expectation should be around 99.5%, and anything greater than 99.9% is very good (spitting out a lot of nines)!

b) **What are the provider's services regarding disaster prevention, recovery, and business continuity?**

Unfortunately, bad things can still happen in this world. Fires, earthquakes, gas explosions, plane crashes, terrorist attacks, power disruptions from utilities, etc., are just a few of the catastrophic occurrences that can take the processes and operations of a business down. The consumer should examine what the provider offers or expects to do when one of these events happens.

- The consumer should determine if the provider has multiple operations centers, capable of delivering the desired service levels, where redundant operations could be located in distributed locales.

- The consumer should seek assurances that there will be backup servers always available in reserve to prevent disruptions of service when the original server fails.

- The consumer should confirm that the provider applies the appropriate performance monitoring to ensure that issues can be recognized and addressed in the most optimal fashion.

c) **What scope of parameters does the provider cover in defining its SLA's (service level agreements)?**

The provider and consumer must clearly define the metrics that will be employed in establishing and adhering to the expected levels of performance. Some of the more common ones are:

- Availability % describes the ability of a service to perform its agreed function when required.

  = (The Agreed Service Time less Downtime) divided by The Agreed Service Time

- Reliability is a measure of how long a service can perform its agreed function without disruption.

  MTBSI or mean time between service incidents.

= Available Time divided by Number of Service Disruptions within this Time

MTBF is also known as Uptime; mean time between failures.

= (Available Time less Total Downtime) divided by Number of Service Disruptions within this Time

Maintainability is a measure of how quickly and effectively a service can be restored to agreed level of operations after a failure.

MTRS is also known as Downtime; mean time to restore service.

(Total Downtime in a Period) divided by The Number of Service Disruptions in that Period

d) What are the compensatory expectations if the SLA's requirements are not met?

The consumer should determine what the provider will offer as compensation to the consumer when the SLA's requirements are not met. These terms may be specific to a first or second occurrence within a month, progressively more painful. They may be expressed as refunds, penalty credits, waived subscription fees, or discounts. The consumer should establish what is expected after SLA failures.

4. Computing Resource Optimization

The consumer should determine the provider's resource management solutions.

a) Does the provider deliver resource scalability, elasticity and deployment automation?

The consumer's usage patterns may be extremely flat, but they are more likely to involve the deployment of resources for applications that demand scalability with elasticity. There may also be specific projects with infrequent high compute or storage demands. The provider's solutions should provide for roles and rules that permit transparent automatic provisioning and releasing of resources, preferably without any form of disruption to the consumer.

b) What tenancy scenarios does the provider support?

The consumer should determine how the provider's solution supports a multiple tenant type of demand.

- Support that is limited to a single tenant per instance of the technology stack is a very expensive solution for the provider, and this is especially true if the stack contains significant licensing costs instead of free-use open source components. The provider's costs can only go up, which means that the consumer's subscription price can only go up.

- Customized multiple instance solutions are still expensive, because there are only so many consumers that the provider can support under this model.

- Multi-tenancy support, especially with load balancing of virtualized resources, is the most cost effective solution for the provider and consumer. The offsetting concerns are around security challenges associated with multiple consumers' data and processes co-existing on a partitioned single instance stack.

c) Are there any device or platform dependencies?

The consumer should identify any dependencies in the provider's solutions that directly or indirectly lock-in to devices or platforms.

5. End-to-end Network Support

Network services should provide the reliable, scalable and efficient interconnection possibilities between all other Cloud service providers on the network to satisfy the consumer's business needs. The consumer's choices about network services are typically independent from the choices about Infrastructure, Platform and Applications providers. They should be abstracted from the MUXs, cables, routers, switches, network management solutions and servers, and focused on Cloud solution optimization.

a) How does the Network Service Provider (NSP) employ traditional web services?

The NSP may limit its role to strictly providing bandwidth at an acceptable level of performance. Or it may provide many of the components and features that are typically provided by other service providers of IaaS, PaaS, and SaaS. The consumer should have a complete understanding of the Cloud portfolio of the NSP and how its various capabilities might provide more options or other forms of risk-mitigation.

b) Does the network emphasize end-to-end performance measurement and assurance?

The network service provider must emphasize a dedication to the performance of the Cloud solution as an end-to-end challenge, and not just an application, computation, or storage issue that they cannot resolve. The end-to-end nature of SLA's that cover KPIs across the data center, the network, to the consumer client should reflect the entire business process preferably with a single point of accountability.

c) How does the network provide scalable and elastic bandwidth to the consumer?

Network spikes are expected to happen in the case of public events, disasters, campaigns, large file transfers, consumer-internal multimedia events, and seasonal application loads. The consumer should be able to determine the scope and scale of network bandwidth scalability and elasticity, and how they will affect responsiveness and latency challenges.

d) What are the network-based limitations associated with disaster recovery?

High volumes compressed into narrow timeframes are a significant challenge with some if not most recovery and continuity scenarios. The consumer should be aware of the limitations that the network will place on these operations, and make the appropriate arrangements that ensure its expectations are met by the overall solutions.

e) What does the provider's network architecture provide?

The consumer should examine the topologies of the various network constructs to determine the range of network possibilities. Latency challenges in access to data stores or movement of data between clients and servers throughout the network must be exposed. Data center location limitations should be considered.

f) Are there limitations in upgrade or virtualization of the infrastructure?

Upgrades to infrastructures occur often, and virtualization of the infrastructure's computing resources are increasingly performed dynamically at runtime. The consumer should be aware of any limitations that the network services will place on the other components of the infrastructure.

6. The service provider's partners and consumer ecosystem

a) Does the service provider create, maintain, and/or host an ecosystem?

The breadth and depth of the infrastructure services portfolio can become so complete and rich that there is an ecosystem formed consisting of various partnering service providers and the service consumers. This tends to serve both the providers and the consumers because it enables context for the complete service portfolio across all providers and provides a forum for consumers to share and drive the service providers' priorities. The consumer must safeguard against lock-in in this situation as well.

H. Cloud IaaS Vendors

The following are some of the leading Cloud IaaS vendors in random order:

- Amazon: Infrastructure Services, Network Services

- IBM: Cloudburst, Tivoli Federated Cloud, Network Services

- AT&T: Cloud Services, Network Services

Shapes In The Cloud

- HP: Cloud Assure, Network Services

- Worth watching: CA (including 3Tera)

- Interesting: Alliance of VMWare, Cisco , EMC

10

The Platform Layer

Providers, consumers, services

Discussion Areas in This Chapter

- *Software platforms*

- *PaaS*

- *Features and technologies*

The Platform Layer

The next area of focus within the framework is the platform layer. Platforms are deployed on top of infrastructure to provide the development frameworks, programming languages and tools enabling the design-time creation and runtime execution of applications. The platform consumer does not manage or control the underlying network, servers, operating systems, or storage of the infrastructure, but has direct control over the deployed applications.

A. Software Platforms

A general definition of a platform that applies to both hardware and software emphasizes its use as an architectural foundation or base. The software platform facilitates the rapid development, management and run-time utilization of computer-based applications.

The simplest and most basic definition of a software platform is that it is a base used to launch software. Basically, software platforms provide the base functionality to communicate bi-directionally back and forth with other software. Given this definition, an operating system is an example of a software platform because applications must interface with them. Some of the operating systems that are common or critical to the success of Cloud Computing include:

- Windows 7
- Mac
- Linux
- AUX, Unix
- Solaris
- Chrome
- MS Windows Azure
- Mobile OS – iOS, Android, Windows Phone 7

The Platform Layer

An application may also be a platform if it can be used as a base to call or launch other software. Web browsers can launch third-party plug-ins to add functionality, and e-mail, calendaring and other client programs can be launched from messaging or other forms of collaboration software.

There are a number of software applications that can be platform-versatile, but never platform-independent. Java is one of the better known examples of a software language, but its versatility is owed in large part to the fact that it is both a language and a platform (JavaEE). A software application is by definition platform-dependent, and each platform assures consistency in any application that is developed using its tools, including for example an SDK for the appropriate web application framework effectively integrated within a development environment's (IDE) UI, text editor, debugger, file management, etc.

Beyond the design and development of applications, each platform supports the runtime launch and execution of its applications' end-to-end processes within the scope of its application program interface (API) library.

B. PaaS: Platform-as-a-Service

Estimates for PaaS revenues and growth reflect increasing Cloud emphasis. Projections for 2013 PaaS service provider's revenues range from $4 to $6 B., 12 to 16% of the global Cloud Computing revenues.

The on-demand nature of IaaS solutions is recognized for the shift of capital costs to operational costs. Cloud Computing platforms are attractive because they let businesses quickly access private and public resources on-demand without the complexities and time associated with the purchase, installation, configuration and deployment of traditional physical infrastructure. A Cloud Computing platform enables its applications to be hosted in an Internet-accessible virtual environment that supplies the necessary hardware, software, network, and storage capacities with assured security and reliability. Since the PaaS provider delivers the platform over a network, the PaaS consumer typically can access the platform using a browser, with no need of other software downloads.

Therefore, the on-demand nature of PaaS solutions reduces costs across the development, deployment and runtime management aspects of the application lifecycle. The consumer doesn't need to invest substantial initial funds to have a development foundation for developers, but pays only for the time, resources, and capacity it uses while scaling up to accommodate its changing business needs. PaaS solutions have empowered large, medium and small companies or entrepreneurial individual developers to launch their own application software delivered as a service (SaaS) to support extensive business processes, including integration of SaaS applications and/or on-premise applications.

1. The PaaS consumer

PaaS environments enable their ISV (independent software vendor) or other consumers' developers and architects through services and API's. Developers then create specific application API's and service-based functions, deploy these applications, configure the application and technology components, integrate applications running in the Cloud and applications running on local client systems, and apply continual service improvement processes to manage system change.

A PaaS solution allows the consumer to deploy applications using programming languages and tools supported by the provider into a virtually infinite reservoir of available computing resources, and thereby substantially reduce barriers from deployment challenges and infrastructure configuration complexity. The consumer probably cannot manage or control the underlying Cloud infrastructure's network, servers, operating systems, or storage, as the platform provides control over its deployed applications and any application environment configurations.

The platform provides well-formed distributed caching, queuing and messaging, workload management, file and data storage, user identity, analytics, etc., and thereby eliminates the platform consumer's need for many disparate components, substantially reducing their SaaS applications' time-to-market.

2. The PaaS provider

A PaaS platform provider integrates a Cloud OS, a development environment, platform middleware, and application software provided to a consumer. A Cloud software platform packages and applies both role- and rule-based access to expose its functional capabilities as loosely coupled services through published API's.

PaaS runtime frameworks execute code according to configuration set by the application owner and Cloud provider. PaaS runtime frameworks can use traditional application runtimes, 4GL, 5GL, and hardened integration ports for easily pluggable external application runtimes.

Cloud platforms employ a higher level of abstraction than most other types of software application platforms, realized through more extensive automation and more robust metadata solutions that deliver an elegant user experience.

A Cloud application platform will likely map to one of the following definition scenarios:

- A software application platform supporting its single SaaS application, run exclusively on a limited infrastructure stack that has been defined by the SaaS application provider.

- The platform used by salesforce.com for its CRM application suite prior to the release of Force.com was an example.

- Note that consumers are probably locked into the infrastructure options of the SaaS provider.

- A software application platform offered as a service (PaaS), supporting 2 or more applications, and run exclusively on a limited infrastructure stack that has been defined by the PaaS platform provider.

  - Google AppEngine and Chrome are Cloud OS platforms, where proficient Java & Python (etc.) developers can deliver Cloud applications.

  - MS Azure is another Cloud OS platform, whereby .Net developers can easily move up to the Azure and create SaaS applications.

  - Force.com used its proprietary APEX language, while Force2.com apparently moves toward a Java-based solution. A metadata-driven solution with substantial support for both developers and non-developers applications, a huge SaaS enterprise ecosystem for innovation and co-marketing, and a strong ERP business presence.

  - OrangeScape and Wolf PaaS are examples of more 5GL (5th generation language) middleware-centric platforms that (like Force2.com) automate many standard software development tasks to simplify application development, enabling non-developers such as SME (subject matter experts) business analysts with domain knowledge to deliver SaaS solutions.

- A software application platform offered as a service (PaaS), supporting 2 or more applications, run on infrastructure stacks as prescribed by the consumer. Service consumers' infrastructure options are more open based on the enablement delivered by the platform provider.

CA 3Tera AppLogic appears to be one example of this solution as a "turnkey" Cloud Computing platform for composing, running and scaling distributed applications, using:

- o Advanced virtualization technologies to be completely compatible with billions of lines of existing infrastructure software code covering fire walls, load balancers, network configurations and database servers, middleware and application code.

- o Operations on the logical structure of the application, enabling packaging of N-tier applications as a single logical system.

- o Software-to-hardware binding that enables applications to be replicated on demand without code modifications.

3. Primary business candidates for PaaS

Some types of platform applications are better candidates than others for earlier Cloud Computing adoption. The predominant focus of PaaS Is support for Cloud applications. Therefore, the primary application of PaaS will be associated with the SaaS applications.

Application development requires substantial investments in platform and infrastructure to support all of the various possible application constructs, configurations and test cases that must be validated. An extensive discussion of these types of demands is provided in a separate section of this book (Use cases for software provider's internal operations).

Platforms are likely required for the same software applications listed above for IaaS (Primary business candidates for IaaS).

C. Key PaaS Evaluation Attributes

Cloud platforms exist primarily to allow quicker development and deployment of Cloud applications at reduced costs. The platform delivers access and control over the deployed Cloud applications (SaaS) and possible application environment configurations and customizations.

Competing PaaS platform offerings can be differentiated by the evaluation of the capabilities they deliver as Cloud services.

1. Platform features

a) Does the platform have the desired features?

- Is the solution a complete technology platform for SaaS ISVs or private developers?

- Is there a library of ready-to-use components to quickly and easily build new features and applications?

- Is it easy to change existing applications?

- Are the required development skills readily available? Long-term?

- Are developers required or can domain-expert analysts succeed in using the platform?

- Are the preferred SDK and IDE solutions available to the platform?

- Are there any elements or aspects of web application frameworks that are limited?

b) What tenancy scenarios does the provider support?

The consumer should determine how the provider's solution supports a multiple tenant type of demand.

- Is the platform limited to single-instance support?

Support that is limited to a single tenant per instance of the technology stack is a very expensive solution for the provider, and this is especially true if the stack contains significant licensing costs instead of free-use open source components. The provider's costs can only go up, which means that the consumer's subscription price can only go up.

- Is the platform limited to multi-instance support?

Customized multiple instance solutions are still expensive, because there are only so many consumers that the provider can support under this model.

- Does the platform support running multi-tenancy applications?

Multi-tenancy support, especially with load balancing of virtualized resources, is the most cost effective solution for the provider and consumer. The offsetting concerns are around security challenges associated with multiple consumers' data and processes co-existing on a partitioned single instance stack.

c) Are there any performance limitations to the platform or its applications?

- Does the platform scale with elasticity in response to demand volatility?

- What are the platform's availability, reliability and maintainability benchmarks?

d) Does the platform have deployment flexibility?

- What platform features may require a lock-in to the platform provider or the provider's network of partners?

- Are there any portability limitations for the platforms data stores, etc.?

- Can the platform be deployed on the consumer's preferred private, public, or hybrid infrastructure?

- Are there any deployment model migration limitations, moving from public to private to hybrid?

- Are there any multi-Cloud application deployment limitations?

- Can the platform consumer and users focus on application development not infrastructure?

- Are there any dependencies in the provider's solutions that directly or indirectly lock-in to devices?

e) Can the platform consumer immediately build applications?

- Will the platform users be able to begin immediately in their development and deployment of applications with absolutely no startup costs required?

- Will the platform be limited to on-demand subscription costs, and minimize the consumer headcount and related costs to support the consumer application solutions?

2. Core platform technologies

a) Does the platform provider employ best standards in Cloud technology?

- Does the platform use a service-oriented architecture?

- How does the platform use object-based and relational-based data models?

- Does the platform employ extensible markup languages throughout its communication and file structures?

- Does the platform support client-server N-tier architecture?

b) Does the user interface apply best web and Cloud design practices?

- Does the interface represent an aesthetic, minimalist and sensible design?

- Is the interface simple as possible, easily map to the real world process, and task-focused?

- Does the interface provide / follow standards for consistency, similarity, and predictability?

- Is there sufficient visibility that establishes a first step and ongoing status toward goal?

- Is there a good balance between control and user freedom in how to use the system?

- Is the interface tolerant, preventing errors where possible and helping users recognize, diagnose, and recover when prevention is not possible?

- Does the interface emphasize recognition over recall, with timely clarity in system response?

- Is there WYSIWYG reporting?

c) Does the platform enable application customization?

- Is there a platform object layer that is open for review and modification?

- Does the platform use Wizard-based tools to enable application development?

- Does the platform support creating custom screens and fields?

- Are there field filtering, sorting and layout tools available?

d) Is the platform's system administration efficient?

- Is the SA function centralized across the platform?

- Are system monitoring and issue trouble-shooting interfaces integrated?

- Are system integrations easy to implement and maintain?

- Are platform release upgrades transparent to the consumer?

e) Is the platform secure?

- Are all elements properly secured to ensure data, models, code, and objects cannot be accessed and modified without proper authentication?

- Does the platform enable its applications to be run in optimal secured modes?

f) What are the platform-based limitations associated with disaster recovery?

The consumer should be aware of the limitations that the platform may place on its design-time or run-time operations, and make the appropriate arrangements that ensure its expectations are met by the overall solutions.

3. The platform's ecosystem

a) Does the platform provider create, maintain, and/or host an ecosystem of partners and platform application developers?

- Does the ecosystem support collaboration, integration and/or a business network between different applications created on the platform?

- Does the ecosystem provide a ready-made sales and distribution channel for applications created on the platform?

- Does the ecosystem provide dedicated platform initialization, configuration and support?

- Does the ecosystem provide a developers' community for sharing tips, tricks, code, etc.?

- Does the ecosystem provide go-to-market, campaigns, and joint-selling programs?

D. Cloud PaaS Vendors

The following are some of the leading Cloud PaaS vendors in random order:

- IBM Cloudburst

Shapes In The Cloud

- Force.com

- Amazon AWS

- Microsoft Azure

- Google AppEngine

- Watch for: Rackspace Platform Services

- Interesting: NetSuite SuiteCloud

11

The Application Layer

Providers, consumers, services

Discussion Areas in This Chapter

- *SaaS*

- *Candidates*

- *Evaluation attributes*

The Application Layer

Application software may be referred to as an "application" or an "app". An application is an executable software unit or grouping of units that are deployed to deliver actions - create, read, update, delete, move, save, calculate or display, etc. - associated with digital information. An application is initiated either by another software unit or by a user for a specific purpose and results in the realization of some benefit for the calling system or user.

A. SaaS: Software-as-a-Service

Estimates for SaaS revenues and growth reflect increasing Cloud demands. Projections for 2013 SaaS service provider revenues range from $26 to $30 B., 60 to 65% of global Cloud Computing revenues.

This layer provides applications built for Cloud Computing, exposing Web interfaces and Web Services for end users, enabling multi-tenant hosting models, connecting disparate systems, and leveraging Cloud storage.

- The SaaS capability provided to the application consumer is to use the provider's applications running on a Cloud platform and infrastructure. The Cloud application consumer may have limited configuration and even some customization capabilities, but it probably does not manage or control the underlying Cloud platform or infrastructure network, servers, operating systems, storage, etc.

- The consumer of a SaaS application is unaware of the physical computing resources - the platform instance or component instances that were applied, the infrastructure's specific servers or storage devices employed, or the physical network components - that are used in performing the application's activities and tasks.

The Application Layer

- SaaS applications are predominantly run on a shared public Cloud hosted by the SaaS provider, but some SaaS applications might be run on a private Cloud exclusively for the application consumer's users. The applications are accessible from various client devices typically through a thin client interface such as a web browser (e.g., web-based email).

- They are based on a service-oriented architecture that encompasses a library of one or more granular software units packaged as individual services that are exposed through one or more API's. The individual services are stateless, loosely coupled and independent. An application orchestration layer is used to combine services into other services, etc., in a manner that delivers one or more complete functions.

- Cloud on-demand applications cost less than on-premise applications, because you don't need to need to pay for all the products, people, and facilities to run them. They are by definition more elastic and scalable, probably more secure, and definitely more reliable than most on-premise applications.

- There are no application maintenance fees, and periodic upgrades are taken care of by the provider on the production code branch, so the consumer's applications automatically have new features and the latest security and performance enhancements, thereby increasing the speed and agility of deployment.

- SaaS applications are either free to the consumer, charged through a blanket per-period subscription fee, a metered set of usage-based charges, or a combination of these.

B. Primary business candidates for SaaS

There are some application candidates that are more likely to be delivered as SaaS applications. The best SaaS candidates are wherever one or more of the following conditions exist:

- There has been a very high level of standardization, key process options, and the definition of best business practices for the application.

- Isolation of the application produces one or more key business benefits without significant transition costs or substantial business risk.

- The application is much better suited for on-demand delivery than the alternatives.

- Competitive pressures may be forcing the application's transition to the Cloud.

Email and collaboration, CRM (Customer Relationship Management), HR (Human Resources), accounting, finance and control, as well as custom-built Cloud applications are well-recognized candidates for SaaS applications. These applications are often up and running in days, whereas more traditional on-premise business software have implementations that can range between 3 and 18 months.

C. Key SaaS Evaluation Attributes

SaaS applications are one of the most visible manifestations of the Cloud's powerful drivers and benefits for most of the people who will be exposed to Cloud Computing solutions. The platform and infrastructure layers as well as the security and management components are fairly invisible to the consumer and its users (unless and until something goes wrong).

There are several <u>provider-centric</u> (Consumer Evaluation of the Provider) and <u>general software-related</u> (Evaluation Based on Cloud Computing Framework) evaluation attributes discussed earlier, so they will not be repeated here. Many of the attributes for SaaS and PaaS are shared, so there is some degree of repetition here.

1. SaaS features

a) Does the scope of the provider's features meet the consumer's expectations?

Early versions of applications are delivering core features that have the greatest value to the initial target market segments and user personas. Cloud features are also likely to initially be the most standard, ubiquitous and best-practiced process and activities. The consumer should determine how the current and projected application features and functions map to their expected application feature and function needs. The consumer should pursue unfavorable gaps with the provider to set expectations for enhancements to meet acceptable minimum scope.

b) Do the capabilities of each feature set in the scope meet the required needs?

Once the scope of features and functions is found to be acceptable, the consumer needs to determine whether the capabilities and functionality within each feature set will meet the requirements of the consumer's users and processes.

2. Single- and multi-instance versus Multi-Tenancy Support

The consumer should determine how the provider's solution supports multiple tenants

a) Is the application limited to single-instance support?

Support that is limited to a single tenant per instance of the technology stack is a very expensive solution for the provider, and this is especially true if the stack contains significant licensing costs instead of free-use open source components. The provider's costs can only go up, which means that the consumer's subscription price can only go up.

b) Is the application limited to multi-instance support?

Customized multiple instance solutions are still expensive, because there are only so many consumers that the provider can support under this model.

c) Does the application support running multi-tenancy applications?

Multi-tenancy support, especially with load balancing of virtualized resources, is the most cost effective solution for the provider and consumer. The offsetting concerns are around security challenges associated with multiple consumers' data and processes co-existing on a partitioned single instance stack.

3. Performance limitations

- Does the application scale with elasticity in response to demand volatility?

- What are the application's availability, reliability and maintainability benchmarks?

4. **Application deployment flexibility**

- What application features may require a lock-in to the provider platform / infrastructure layers, or to the provider's network of partners?

- Are there any portability limitations for the application's data stores, etc.?

- Can the application be deployed on the consumer's preferred private, public, or hybrid infrastructure?

- Are there any application deployment migration limitations, moving from public to private to hybrid?

- Are there any multi-Cloud application deployment limitations?

- Can the consumer and users focus on doing their business and not be concerned with any platform or infrastructure issues?

- Are there any dependencies in the provider's solutions that directly or indirectly lock-in to devices?

5. **Core application technologies**

a) **Does the application provider employ the best standards in Cloud technology?**

- How does the application use object-based and relational-based data models?

- Does the application support client-server N-tier architecture?

b) Does the user interface apply best web and Cloud design practices?

- Does the interface represent an aesthetic, minimalist and sensible design?

- Is the interface simple as possible, easily map to the real world process, and task-focused?

- Does the interface provide / follow standards for consistency, similarity, and predictability?

- Is there sufficient visibility that establishes a first step and ongoing status toward goal?

- Is there a good balance between control and user freedom in how to use the system?

- Is the interface tolerant, preventing errors where possible and helping users recognize, diagnose, and recover when prevention is not possible?

- Does the interface emphasize recognition over recall, with timely clarity in system response?

- Is there WYSIWYG reporting?

c) Does the platform enable application customization?

- Is there an application business object layer that is open for review and modification?

- Does the application support creating custom screens and fields?

- Are there field filtering, sorting and layout tools available?

d) Is the application's system administration efficient?

- Are system monitoring and issue trouble-shooting interfaces integrated?

- Are system integrations easy to implement and maintain?

- Are application release upgrades transparent to the consumer?

e) Is the application secure?

- Are all elements properly secured to ensure data, models, code, and objects cannot be accessed and modified without proper authentication?

- Does the application run in optimal security?

f) What are the application-based limitations associated with disaster recovery?

The consumer should be aware of the limitations that the application may place on its operations, and make the appropriate arrangements that ensure its expectations are met by the overall solutions.

6. The application's ecosystem

a) Does the application provider create, maintain, and/or host an ecosystem of partners and application consumers?

- Does the ecosystem provide dedicated application initialization, configuration and support?

- Does the ecosystem provide a developers' community for sharing tips, tricks, code, etc. for customization?

D. Cloud SaaS Vendors

The following are some of the leading Cloud SaaS vendors in random order:

- Salesforce.com/AppExchange

- Google Apps

- Microsoft Live Services/BPOS

- Microsoft Portfolio

- Ariba

- Watch for: BMC Remedy OnDemand

- Interesting: NetSuite

12

Cloud Security Considerations

Challenges, standards, evaluations

Discussion Areas in This Chapter

- *Access, data retention*

- *Portability*

- *Cloud Security Alliance domains*

Cloud Security Considerations

Cloud technologies and architectures achieve many of the Cloud's benefits through pooling of physical computing resources that are shared "virtually" or with multiple tenants within a single instance. The consumers and their users need to be able to trust that their data and applications are secure from others, and that access is provided only to authorized users.

A. Primary Cloud Security Challenges

There are several areas of primary challenges for Cloud security that must be met, and there are a variety of ways that these challenges can be recognized and dealt with.

1. Basic Cloud security concerns

The following security issues are common to most discussions of Cloud Computing security.

- **Access to external data and processes** introduces new risks to the consumer when its data and applications are processed outside of its in-house "four walls". The controls that it once defined and executed must now be defined within a new context and executed as a part of the delivered services from the provider. Servers must be able to authenticate remote users and remote servers requesting services or applications. An inability to authenticate leads directly to unauthorized users granted access to data and applications.

  Authentication for access can be based on three points: something the user knows, something the user is, and/or something the user possesses. Providers and consumers should utilize an authentication solution that requires at least two of these three factors to be given priori to access. This is commonly referred to as a 2-factor authentication practice.

Data security breaches are expected to increase in 2011, correlated to increased Cloud services activities and social media applications in general. For example, personally identifiable information (PII) breaches can lead to substantial monetary damages for first and third-party expense issues unless covered by some form of insurance for information privacy and security. Standard PII might include:

o   Full name, maiden name, mother's maiden name, or aliases.

o   Social security numbers (SSN), passport numbers, driver's license numbers, taxpayer identification numbers, financial account or credit card numbers.

o   Street addresses, email addresses

- **Compliance with laws** regarding the integrity of the consumer's data resides with the consumer, even when it is being processed and managed by a service provider. Certifications and compliance audits are critical to the consumer's trust and confidence in the provider's ability to deliver compliant performance. The consumer's business is responsible for any havoc caused through its providers' operations, even if it did not cause the damages.

- **Privacy restrictions** require special processing and storage of data within appropriate legal jurisdictions according to the consumer's locations and the trading partners of the consumer.

- **Data storage and handling within shared resource situations** employing virtualization and multi-tenant technologies is a major area of concern within Cloud Computing solutions.

- **Data Recovery and business continuity practices** become more critical with Cloud Computing, when the provider may or may not provide these types of services in the manner that the consumer might expect to sustain its business through a disaster.

- The ability for the consumer to have **on-demand audit access** as needed is a serious challenge for co-located and cross-data center applications.

- The **ownership and accessibility** by the consumer of its data must not be lost regardless of what may happen to the ownership and performance of the service provider's assets and processes.

2. Convergent database and event management security

To gain a comprehensive view of the activity in its database and Cloud environment, a consumer will need to feed its Database activity monitoring (DAM) information into its IT security information and event management (SIEM) tool. Database information can be combined with event management visibility on network traffic, server configurations, data exfiltration attempts, user activity, etc. to find attack and security gap patterns.

This challenge is exacerbated by the virtualization and tenancy technologies prevalent within Cloud Computing, where requesting servers may not provide adequate authentication. Tying DAM information into the SIEM allows an organization to more easily correlate the activity a user might have done on a front-end application with the query activity by an application server sent directly into the database.

3. Cloud security standards organizations

Due to the significant challenge to Cloud Computing adoption that security issues present, there are several organizations that are busy at work on establishing standards and protocols for Cloud security.

The Cloud Security Alliance recently published Version 1.1 of its Cloud Controls Matrix (CCM). This control document is expected to provide the fundamental security principles to guide Cloud service providers and any prospective Cloud consumers in assessing the overall security risk of a Cloud provider.

The CCM provides a detailed matrix mapping of its security standards expectations aligned with the foundations of the other industry-accepted security standards, regulations, and controls domain frameworks (see below), augmenting internal control direction for SAS 70 references provided by Cloud providers. The CCM framework provides providers and consumers with common structure, detail and clarity regarding Cloud security.

The CSA CCM framework provides:

- 11 major security control areas with 98 identified control standard specifications.

  o Compliance (6 security standards defined)

  o Data governance (8 security standards defined)

  o Facility security (8 security standards defined)

  o Human Resources security (3 security standards defined)

  o Information security (34 security standards defined)

  o Legal (2 security standards defined)

- o Operations management (4 security standards defined)

- o Risk management (5 security standards defined)

- o Development release management (5 security standards defined)

- o Resiliency (8 security standards defined)

- o Security architecture (15 security standards defined)

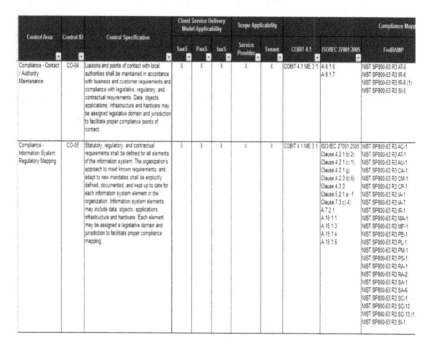

Figure 12-1 a snapshot of Cloud Security Alliance's Cloud Controls Matrix 1.1

- Applicability of these security control standards to 13 Cloud-related domains

- o Delivery models affected (3): IaaS, PaaS, SaaS

- o Scope role (2): Provider, Tenant (Consumer)

o Security compliance domains, cross-reference mapping of the CSA CCM standard to the corresponding standard specification for another security standards domain (the 8 mapped-to domains are named here to allow the reader to research if desired).

- ISACA COBIT 4.1

- HIPAA / HITECH Act

- ISO/IEC 27001-2005

- NIST SP800-53

- FedRAMP

- PCI DSS v2.0

- BITS Shared Assessments , AUP v5.0 / SIG v6.0

- GAPP (August 2009)

B. Key Security Evaluation Attributes

Security is the biggest issue for most prospective consumers considering the move to a Cloud Computing model. The primary fears are associated with placing their core critical data outside of their physical domain and exposing their mission-critical processes to the control of others.

1. Access and data control considerations

a) Does the provider guarantee that only consumer-authorized users have:

- Is there a 2-factor authentication required for access?

- Access to read and copy critical data?

- Access to change or delete critical data?

- Access to perform mission-critical business processes?

- Access to change or remove mission critical processes' configuration and customization?

b) How does the provider define security responsibility and liability?

The provider should have or implement appropriate measures to ensure the security of Customer data. It should commit to a well-formed definition of its responsibility for assuring consumer data security, and it should assume liability under that limited definition for any loss of the consumer's data. The consumer probably should have the responsibility and associated liabilities for its "client" security scope, e.g. user passwords.

c) What protection mechanisms and techniques are utilized in your offerings?

- Firewalls?

- Full-time monitoring?

- Intrusion detection?

- SSL and application security?

d) What 3rd party security certifications practices has the provider received?

- SAS70 (Statement on Auditing Standards Number 70) Type 1, regarding audited fair description of controls and their suitability?

- SAS70 Type 2, named controls effectively operating as expected?

e) What independent standards do the provider's offerings support and enable?

- CSA, NIST, and other?

- ISO 27000 (International Standards Organization) regarding security techniques and ISACA (Information Systems Audit and Control Association) for management systems overview and vocabulary?

- ISO2000, ITIL v3 (Information Technology Infrastructure Library version 3) ITSM (Service Management), and/or COBIT (Control Objectives for Information and related Technology) regarding best practices and integrated processes?

2. Data retention

When the consumer's data is controlled within its own "4 walls", it does not have to worry too much about data rights retentions. However, when the consumer's data is to be processed, stored and managed by a Cloud provider, the consumer is faced with new challenges in this area.

a) Does the consumer retain exclusive ownership rights to their data?

The service provider's role and limitations regarding the consumer's data should be clearly defined. The data must continue to be exclusively owned by the consumer regardless of the abstractions associated with the physical resources employed by the service provider. Further, the privacy laws that apply to the consumer and its data must be upheld and enforced by the service provider's operations associated with the services it is providing to the consumer.

b) What are any other expected uses of its data by the provider?

The consumer needs to understand in clear terms what other uses of its data by the provider are expected to be acceptable. Perhaps the data must be used to establish volume expectations, or monitor categories of usage, or to test development changes and additions for "real-world" applications. The consumer should fully understand what the provider is thinking is acceptable and make the appropriate limitations.

c) Does the provider guarantee that the consumer's data shall never be shared with or sold to others?

This may seem a bit redundant with the question above, but it isn't. The provider may be expecting to be able to share the consumer's data with its partners, or that it can sell the data for research information to 3rd parties. The consumer must expressly limit this type of data integrity abuse.

d) What are the provider's policies regarding the portability of the consumer's data?

The consumer should determine what the provider's abilities and policies are regarding the portability of the consumer's data. Basically, if the data is extremely difficult and costly to move, this may translate into the consumer being "locked in" to the provider or limited by the provider.

Despite the best efforts of the service provider and the consumer to make their Cloud relationship work, things may not work out. The consumer or the provider may cease business operations, or may just decide to discontinue the relationship. This may happen because the consumer wants to bring the computing resources it is using back under its own physical direction or to take its business to another service provider. Another possibility affected by the lack of portability is the consumer's expectations regarding data recovery and continuity.

The consumer should clearly establish the expected data portability solutions and policies before committing to the provider's services.

e) Does the consumer have full and complete access to its data for creating, reading, updating, deleting, transferring to other applications, and downloading in expected formats without any limitations?

The service provider and consumer must clearly establish the service level expectations around the data. The consumer should have "unlimited access" 24/7/365 to its data and its data processes, to be able to perform any CRUD activity, to transfer its data to or from other applications (e.g. using the provider's API's), and to downloading its data in expected formats as desired.

The provider may have certain situations when it must be able to have reasonable periods of blocked access to the consumer, such as rolling in new code updates or releases. But the provider should be able to schedule these events in advance and provide advanced notice for the consumer. There should be other options that enable non-stop processing, such as using alternate or backup resources for limited periods prior to full migration.

3. General Cloud Security Issues

a) How does the provider's service deal with expected public deployment challenges?

In addition to existing enterprise email, antivirus, identity, authentication, and encryption product offerings, consumers must have identified the new challenges and solutions that operating in a hybrid Cloud presents. In a public or hybrid Cloud, the concept of a secure perimeter to a data center building disappears, as applications and data are run in remote locations whose physical security is controlled primarily through SLAs.

b) How does the provider assure security with multi-tenant and virtualized resources?

- Use of the Cloud inherently implies multi-tenancy support from the infrastructure up through the platform and application layers, where data security is only as effective as the partitioned architecture's integrity.
- The infrastructure resources are extensively virtualized, whereby a deployed application is unaware of the applications that were running on the physical hardware before or after it is run. Security is only as effective as the integrity of the resource task completion and complete resource release without any data or process residue, or other exposure.
- The consumer needs to have a comprehensive and clear understanding of the provider's trust services, and the associated monitoring, proactive and reactive event processes.

C. Cloud Security Vendors

The following are some of the leading Cloud Security vendors in random order:

- IBM

- EMC/RSA (incl. Cyota)

- Symantec

- CISCO

- Oracle

- Watch for: Sonoa Systems

- Interesting: Ping Identity

13

Cloud Management Considerations

Design, deployment, management

Discussion Areas in This Chapter

- *Cloud lifecycle management*

- *Solution attributes*

- *Component overview*

Cloud Management Considerations

Cloud Management is defined here as a process supported by a system comprised of a platform, tools, and applications, that directly enable a Service Provider to define, design, develop and create, implement, deploy, maintain and enhance a Cloud that delivers SaaS, PaaS, and/or IaaS services. This chapter introduces the significant components of a Cloud Management solution and how they would be used by a service provider.

The primary considerations for a discussion of Cloud Management solutions are:

- The management of a Cloud throughout its lifecycle.

- Some key attributes of a Cloud Management solution.

- A high-level example of Cloud Management architecture.

A. Cloud Lifecycle Management

Most Cloud Management solutions typically support multiple aspects of a Cloud deployment that can be broken down into separate phases of a lifecycle involving design, deployment, and operational management.

1. Cloud design

Design must follow function here. The service provider must decide several things that will drive the design of the Cloud services it will provide, and then create the Cloud design.

a) Determine the target Cloud service attributes

In many ways, the issues here parallel the evaluation parameters expressed in the earlier chapters regarding the consumer's evaluation of alternate service providers and the services they provide. The service provider should review and consider each of these evaluation attributes in the Cloud services that they offer, and proactively provide its position on each of these attributes as the primary way for it to differentiate itself and its services to prospective consumers.

- What are the business challenges that the Cloud services will resolve? (See "An Evolution")

- What are the positive attributes that the Cloud services will emphasize? (See "Defining the Positive Attributes of the Cloud")

- What factors that might inhibit Cloud adoption should be expected and overcome? (See "Recognizing factors that inhibit Cloud adoption")

- What types of virtualization and/or multi-tenancy features will be provided? (See "Shared Resources Serve Multiple Consumers")

- What Cloud deployment models will be employed to deliver the Cloud services? (See "Cloud Deployment Models")

- Which Cloud Computing framework layer(s) is being offered? (See "The Cloud Computing Framework")

- How will the Cloud services enable the target consumer segments to align strategies? (See "Defining / aligning the consumer's Cloud Computing strategies")

- Will the Cloud services portfolio support the target consumers' Cloud adoption roadmap? (See "Defining the

Roadmap for complete realization of Cloud Computing solutions")

- What ecosystem <u>partners</u> and structures will be required / offered to add greater value? (See "Provider's management of partner providers")

- What are the types of <u>support</u> that will be offered for the Cloud services? (See "Support")

- How will the cloud service subscriptions be <u>priced</u> and measured? (See "Pricing, subscriptions, and metrics")

- What is the expected process for identifying, prioritizing, and implementing <u>enhancements</u>? (See "General Cloud Computing Framework Evaluations")

- How will the Cloud services' <u>integration</u> challenges be met? (See "Integration")

- What are the <u>usability</u> offerings, methods and solutions available to the consumer? (See "Usability")

- For <u>IaaS</u> Cloud service offerings, what are the key attributes of the services? (See "Key IaaS Evaluation Attributes")

- For <u>PaaS</u> Cloud service offerings, what are the key attributes of the services? (See "Key PaaS Evaluation Attributes")

- For <u>SaaS</u> Cloud service offerings, what are the key attributes of the services? (See "Key SaaS Evaluation Attributes")

- How will the consumer's <u>access and data control</u> challenges be resolved? (See "Access and data control considerations")

- How will the consumer's <u>data retention</u> challenges be resolved? (See "Data retention")

- How will the consumer's <u>other security challenges</u> be resolved? (See "General Cloud Security Issues")

- How will the Cloud services' success stories be provided as <u>references</u> to prospects? (See "References")

b) Create the design

The results of the design attributes analysis above will establish the functional use-case requirements and non-functional Cloud service requirements, and lead to the architectural design that best supports the balance of security, efficiency, agility and flexibility. In turn, this will define the object, process, and data models to support the expected service workloads and end-to-end flows.

The Cloud design toolset must place a great emphasis on reusability. The structure of the Cloud should make extensive use of service definition scripts, templates, patterns and virtualization image management to ensure that when Cloud resources have been well-formed into Cloud services, they are easily recreated (cloned, new instances created), modified and implemented as needed. These reusable components of the Cloud service may be constructed quickly from scripts, templates, and OS (guest operating systems) in images to provide expected behavior at various stages of the service delivery, e.g. server boot, operations, and shutdown; or application initialization, configuration, integration, and customization.

2. Cloud deployment

The Cloud service design will make the implementation as easy as possible. The high degree of modularity, standardization, and automation achieved through the realization of the design will substantially benefit the Cloud services' deployment.

The service provider's Cloud services will likely involve infrastructural resources that have been virtualized or software that has been designed to support multi-tenancy. The service provider's Cloud service configuration management system and database should be designed to accommodate most of the key IT service and business service attributes of the individual consumer. This might be using the consumer's static IP address ID when virtualized server resources are spun up on demand; or the integration points and data exchanged in the application's supported processes. There are several related key points of consideration in the deployment process that are covered in the following chapter on ITIL, ITSM, and BSM.

The best case here is that the highly reusable design elements are easily deployed to work seamlessly together across one or multiple Clouds as needed. Entire deployments can be logically cloned and saved for reuse leverage in quickly and easily making similar deployments on-demand. Ideally, resources could be dynamically configured at runtime to produce predictable abstraction, loading, and performance results.

3. Operational management

Refer to the Cloud Management Platform's architecture below for additional references to the Cloud service management functions directly supported by the service provider's operation support services and business support services in the delivery of the offered cloud services.

Operational management includes monitoring and maintaining deployment-wide resources, and launching additional or replacement resources in the proper context. Automation through proactive agents assure that expected availability, resiliency and maintainability performance levels are met and that necessary scalability and elasticity functions are performed without intervention.

B. Attributes of Cloud Management Solutions

There are some attributes that are likely to be present in the process and systems associated with Cloud Management solutions.

- **A comprehensive system with a centralized platform**

  The management of a Cloud from design through deployment and ongoing maintenance requires a system that is easy-to-use and comprehensive in its compatibility coverage of popular technologies and structures. It must be capable of building and deploying a broad spectrum of possible Cloud services and deployment models environments' infrastructure and platform computing resources, delivering Cloud command, control, provisioning and monitoring. The platform should support a landscape of heterogeneous Cloud solutions supporting multiple applications. The platform solution should be infrastructure- and application platform-agnostic, enabling a Cloud service provider the best opportunity to work with their preferences in architecture, programming languages, data stores, and other Cloud solution providers.

- **Portability, Lock-in Avoidance**

The Cloud Management platform should work equally well in single- or multi-Cloud situations with concurrent utilization of multiple application platforms and/or infrastructure solutions within the same landscape. This will:

- Allow freedom to select preferred infrastructure components for SaaS and PaaS solutions, moving away from reliance on vendors' proprietary languages, removing restrictions imposed on the use of data, and increasing visibility into the application's performance.

- Enable the adoption of Cloud solutions under any deployment model, including multi-Cloud, with ease of migration, expansion, and/or addition of others.

- **Quick application start-up**

The solution typically provides a comprehensive and steadily growing library of standard images, scripts, server templates or patterns, solution packages, and building blocks for most popular infrastructure and platform components. A provider can create its own components, employ these or existing standard components to provide initial application architecture, and then make any changes or modifications to provide cloud service deployments that will meet their unique application requirements.

- **Control access**

An effective Cloud management solution will create and manage user access and permissions that provide layered access to production systems and full compliance with corporate governance policies.

- **Test deployment options**

Cloud Management Considerations

There are a number of situations where a Cloud provider would like to test, compare, and evaluate two or more stacks of application architecture, development languages, software platforms, data stores, and supporting providers. Cloud management enables the design and reuse of complete deployments, whereby a key independent variable is changed to determine the implications of each possible application deployment variant.

This supports the determination and selection of one or more optimal infrastructure and platform solutions. It also provides solution pre-release testing for all supported infrastructure and platform solutions. The user can quickly and easily spin-up a complete environment on-demand, and then shutdown the environment when the task is completed.

A user can clone the test environment and move it partially or completely into a staging environment, and ultimately move the staging environment into the production environment. Version control, roll-forward/roll-back, and deployment archiving are intrinsic to enhancements.

- **On-boarding new web-site applications, consumers, users quickly**

Cloud management can maintain archives of highly-configured deployment environments and server templates or patterns for frequent reuse in quick and easy deployments for new web-site applications, consumers, and users, etc.

- **Automation**

Cloud management provides automation solutions for runtime server configuration, problem remediation, and elastic auto-scaling of compute, storage and network resources based on demand and rule sets. Solutions enable the quick launching of resources to meet viral web demands and deploying resources to provide grid computing for massively parallel processing. Each systems administrator can now manage scores of Cloud service deployments and thousands of Cloud servers across the entire deployment lifecycle.

- **Visibility to Track and Control Usage**

Cloud service level agreements and contracts will require Cloud management solutions to provide visibility into the various end-to-end components within a Cloud infrastructure, allowing systems administration to manage, monitor, test, analyze, and re-deploy applications under controls within policies and best practices. Tools include a unified billing report with analytics of daily, weekly and monthly operating costs for computing resources.

For platform services, the objects, logic, process flow and control, and user types that are employed in the application's process activities must be monitored and key business metrics analyzed and reported on.

For application software services, transactions, flows and user accounts will dictate much of the performance management, and business analytics and reporting are intrinsic to the application role in end-to-end management of business processes.

C. Cloud Management Architecture

Cloud Computing is not the first business model involving the provision of on-demand service packages for prearranged usage-based rates and subscription prices. Telecom and financial services are examples of industries that have been service providers for quite a long time. Cloud service providers share some common components of their service provision platforms with these earlier experts. But it should be noted that the advance of IT Service Management practices (see "ITIL, ITSM and BSM Considerations") and Cloud service technologies have blurred some of the more traditional definitions and distinctions.

1. Operations Support Services

An Operations Support System is typically referred to as OSS by service providers. This system is usually a suite of (predominantly) software applications that support back-office or "back-of-office" activities. These systems are employed by service planners, service operations, service architects, service support, and engineering teams to manage the operation of a Cloud service provider's virtualized infrastructure for the provisioning of resources, design and development of services, and the delivery of services.

There are a variety of applications that can be considered in this context, and the key attribute here is that we refer to the operations support services these systems deliver to the Cloud service provider as a part of a Cloud management platform, enabling the Cloud service provider to deliver their Cloud services to the consumers. Figure 13-1 is a representation of some of the operations support services that might be included in the Cloud management platform.

A Service Provider's Operations Services

Operation Support Services: Operation functions for management of Cloud Services		
Service Level Mgmt	Incident Mgmt	IT Controls & Policies
Capacity & Performance	Knowledge Mgmt	Event and Impact Mgmt
Service Templates	Problem Mgmt	Asset & Configuration
Image Management	Identity Mgmt	Provisioning
....	....	....

Figure 13-1 identifies some possible services for operations support

The figure suggests several key operations support IT services that would enable the service provider's management of their Cloud services:

- Managing the details of service level agreement parameters.

- Managing service asset capacity and performance attributes supporting SLA parameters.

- Managing the definition of IT policies and the related controls that ensure service adherence.

- Managing the impact analyses for incidents, problems, or changes affecting business processes, and the definition, monitoring and responses for events.

- Managing the inventories of services assets and the configuration items supporting the Cloud services.

- Managing the provisioning lifecycle from resource identification through allocation and release.

- Managing the service lifecycle's support functions for user/role identity, incidents, problems, and knowledge.

- Managing high levels of reusability and automation through the use of service templates and virtualization images to implement and maintain Cloud services.

- Managing any operations activities that indirectly support the business processes capability to successfully ensure Cloud services are delivered as expected.

2. Business Support Services

A Business Support System is typically an integrated suite of software applications that directly support customer-facing activities, sometimes referred to as front-office or "front-of-office" activities. A BSS may also provide customer-facing interface masks for OSS back-office activities initiated directly by a service assurance contact with the customer.

The key attribute here is that we refer to the business support services these systems deliver to the Cloud service provider as a part of a Cloud management platform, enabling the Cloud service provider to deliver their Cloud services to the consumers. Figure 13-2 is a representation of some of the business support services that might be included in the Cloud management platform.

A Service Provider's Business Services

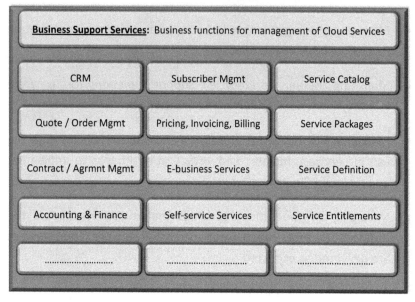

Business Support Services: Business functions for management of Cloud Services

CRM	Subscriber Mgmt	Service Catalog
Quote / Order Mgmt	Pricing, Invoicing, Billing	Service Packages
Contract / Agrmnt Mgmt	E-business Services	Service Definition
Accounting & Finance	Self-service Services	Service Entitlements

Figure 13-2 identifies possible services for business support

The figure suggests several key business-supporting IT services that would enable the service provider's management of their Cloud services:

- Managing the customer relationship's attributes, opportunities, leads, campaigns, service desk, customer support processes, etc.

- Managing the quotation and order management associated with Cloud services.

- Managing the service level agreements and contracts.

- Managing the service pricing, invoicing, and billing activities.

- Managing e-commerce and self-service processes.

Cloud Management Considerations

- Managing the accounting controls and finance analyses, and the business workflows.

- Managing the subscription details for subscribers and subscribed services.

- Managing the comprehensive catalog of services, hierarchies, dependencies, etc.

- Managing the definition and packaging of services, and the service access rights and entitlements

- Managing any business activities that directly support the business processes capability to successfully deliver Cloud services as expected.

3. Dashboards, Analytics, Metrics and Reporting

The Cloud management platform provides the tools to support ongoing management and continual improvement of the Cloud and its services. These tools are provided through IT services which allow the definition of key metrics and indicators, monitoring of actual performance, analyses of trends and options, dashboard consoles for decision support, and standard and ad hoc reporting activities.

- Dashboards will provide a graphical representation of service performance and availability. Results may be updated in real-time and shared across management reports and web highlight pages. They typically would support service level management, event management, performance management, and/or incident management.

- Analytics can be descriptive, predictive and/or prescriptive. Data can be used to model past behaviors to predict expectations and options, and to identify best probabilities of outcomes in balancing capacities and assuring proper availability levels.

- Key metrics can be identified to enable management of a business or operational process, an IT service or some other supporting activity. The most critical metrics can be defined as KPI's (key performance indicators) and used to actively manage and report. Cost, efficiency, and effectiveness are the primary purposes for monitoring KPI's.

- Reporting can be provided as standard scheduled reports or as ad hoc reports that allow access to one-off analyses that suit the current issues and needs.

4. API's and Service Portals

The Cloud service consumer may need to customize or extend the Cloud service functionality to best suit or fit their business process needs. Or the consumer may require specific integrations between the Cloud service and other consumer applications and platforms to create an end-to-end business process.

If the service provider wishes to allow this type of enablement to its service consumers, the Cloud management platform will be required to provide one or more API's (application programming interfaces) to its service consumers. The API's will expose the objects, logic, and processes of the standard Cloud service to the consumer's own developers to enable interaction with other software. The API provides an abstraction context between the consumer's added code and the standard code, allowing changes that won't be deprecated with each standard service upgrade from the provider.

There are at least two possible interfaces for the platform. One is through a portal for the delivery of the cloud service to its consumers, and this portal will have sufficient access for making the API extensions etc. Another interface would be a portal for all service "exchanges". These would include services that have been changed or developed by external service developers, or OEM services from other service providers, that are to be subsumed by one or more of the service provider's offered services.

Figure 13-3 is a representation of the Cloud management platform architecture described above.

Cloud Management Platform

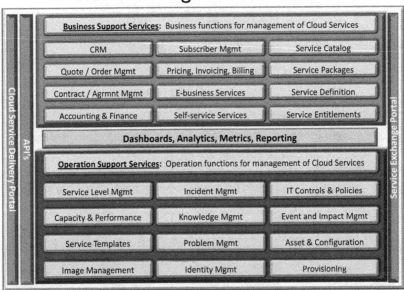

Figure 13-3 represents one example of a Cloud Management Platform

5. A Cloud Management System

A Cloud service provider will require virtualized infrastructure that has been purposed or allocated to host the Cloud service(s) that it is offering as well as the Cloud management platform that has been described above.

Shapes In The Cloud

The overall combination of the Cloud management platform, the Cloud hosting platform, the virtualized infrastructure supporting the hosting platform and the Cloud management platform, as well as the hosted Cloud services, would be defined as a complete Cloud Management System.

Figure 13-4 below represents an example of a complete Cloud Management System architecture.

Cloud Management System

Cloud Services
Applications: Software-as-a-Service (SaaS)
Platforms: Platform-as-a-Service (PaaS)
Infrastructure: Infrastructure-as-a-Service (IaaS)

Virtualized Infrastructure for Cloud and Cloud Management Platforms

Cloud Management Platform

Business Support Services: Business functions for management of Cloud Services

CRM	Subscriber Mgmt	Service Catalog
Quote / Order Mgmt	Pricing, Invoicing, Billing	Service Packages
Contract / Agrmnt Mgmt	E-business Services	Service Definition
Accounting & Finance	Self-service Services	Service Entitlements

Dashboards, Analytics, Metrics, Reporting

Operation Support Services: Operation functions for management of Cloud Services

Service Level Mgmt	Incident Mgmt	IT Controls & Policies
Capacity & Performance	Knowledge Mgmt	Event and Impact Mgmt
Service Templates	Problem Mgmt	Asset & Configuration
Image Management	Identity Mgmt	Provisioning

Cloud Service Delivery Portal — API's — Service Exchange Portal

Figure 13-4 represents one example of a Cloud Management System

D. Key Cloud Management Evaluation Attributes

The best Cloud management solutions will ultimately be vendor-neutral, platform- and infrastructure-provider agnostic. Some Cloud management solutions today come from large Cloud vendors that have a substantial solution footprint across the Cloud Computing framework with a very high capability to manage successful Cloud Computing solutions. They have both a deep and broad spectrum of Cloud Computing solution knowledge, but quite often with a bias towards their own solutions.

The list of attributes discussed below is intended to highlight where vendors are on the path towards providing the best Cloud management solution.

1. Alignment of strategies, financial justification and plan

a) Is the solution capable of full Cloud architecture framing?

- How does the solution develop and analyze architecture alternatives and options?

- Does the solution have integrated economic analysis and justification?

- Does the designer provide point relationships to the policy, compliance, or security drivers?

- Does the designer support a global, standards-driven strategy and planning alignment?

2. Coverage of Complete Cloud Lifecycle Management

a) How does the Cloud management solution support the design phase of the lifecycle?

- Does the solution provide design activities for resource virtualization, VM images, scripts, server patterns or templates, etc.?

- Does the design provide reusable server behavior boot, operation, and shutdown solutions?

b) How does the solution support the operations phase of the lifecycle?

- Does the solution provide complete application deployment?

- Does the deployment include dynamic server configurations at start of runtime?

- Can entire deployments be logically cloned and reused?

c) How does the solution support the performance monitoring phase of the lifecycle?

- Does the solution provide deployment-wide operational performance monitoring?

- What types of proactive and reactive automation features manage scalability and elasticity functions and assure availability, resiliency and maintainability performance levels are met?

3. Cloud management features

a) Does the provider employ a comprehensive system with a central platform?

- Does the platform provide coverage for all popular Cloud deployment models?

- Is it easy-to-use?

- Does it provide all infrastructure and platform computing resources, delivering Cloud command, control, provisioning and monitoring?

b) Does the platform support portability and avoid lock-in?

- Can the platform work equally well under any deployment model, single-or multi-Cloud, with ease of migration and/or expansion to others?

- Does the platform support the mixed-vendor landscapes of applications, platforms and/or infrastructure solutions?

- Does the consumer have freedom to select preferred infrastructure components for SaaS and PaaS solutions, moving away from reliance on vendors' languages, data restrictions, and limited performance visibility?

c) Does the platform support quick application start-ups?

- Does the platform provide a comprehensive and steadily growing library of standard images, scripts, server templates or patterns, solution packages, and building blocks for most popular infrastructure and platform components?

- Does the solution enable extensive component design and archiving?

d) How does the platform assure access integrity?

- Does the platform create and manage user access and permissions to provide layered access?

e) How does the platform support test and deployment options?

- Can the consumer test, compare, and evaluate two or more stacks of application architecture, development languages, software platforms, data stores, and supporting providers?

- Can a consumer clone the test environment and move it partially or completely into a staging environment, and ultimately move the staging environment into the production environment?

- Does the deployment provide version control, roll-forward/roll-back, and deployment archiving?

f) How does the platform support new web-site applications, consumers, and users?

- Does the solution employ archived copies of highly-configured deployment environments and server templates or patterns for frequent reuse scenarios?

g) How does the solution provide resource optimization automation?

- Does the solution provide automation solutions for runtime server configuration, problem remediation, and elastic auto-scaling of compute, storage and network resources?

- Does the solution support resource allocation for viral web demands and massively parallel processing?

- Does the solution permits systems administrator to become substantially more productive?

h) **How does the solution provide deployment visibility to track and control usage?**

- Does the platform cover end-to-end components visibility within a Cloud infrastructure?

- Can the solution allow systems administration to manage, monitor, test, analyze, and re-deploy applications under controls within policies and best practices?

- Are there tools for unified billing/usage analytics providing periodic views?

E. Cloud Management Vendors

Cloud management solutions are where a lot of the heavy lifting for Cloud Computing adoption and sustainability occurs. There will be a tremendous amount of demand for integrated platform solutions that will come together to fully deliver the Cloud management support that the service provider needs.

Every one of these leading or top Cloud management vendors has probably spent billions of dollars in building and/or acquiring its capabilities, and each has access through its installed base to recognize great success. But the more successful ones will aggressively pursue working with and through other Cloud-enabling technology partners that will quickly ramp up their knowledge of how best to deliver the top features and values that Cloud providers and consumers will be looking for. One should expect to see alliances forming at multiple levels and multiple touch points that:

- Provide a portfolio of multiple sets of platforms and applications that support accurately segmented markets, yet build a highly adaptable and extensible model that can be employed across these sets.

- Establish architectural capabilities for increased agility in creating and adopting standards, latest tools and technologies, and extraordinary service-based offerings.

Some of the leading vendors of Cloud Management solutions (in random order) are:

- Force.com

- IBM Cloudburst

- Cisco USD/UCS/ICDP (Integrated Cloud Delivery Platform)

- BMC Cloud Lifecycle Management Platform / ICDP (Integrated Cloud Delivery Platform)

- Rackspace Cloud Servers/Sites/Files

- EMC/VMware

- Amazon S3

- Watch for: RightScale

- Interesting: Novell's Xen Cloud Management

14

ITIL, ITSM, and BSM Considerations

Service lifecycle

Discussion Areas in This Chapter

- *Lifecycle phases, functions, processes*

- *IT management*

- *ITIL and business services*

ITIL, ITSM and BSM Considerations

Cloud Computing is a relatively new business-enabling, service-based IT deployment, delivery and consumption model. Cloud Computing service providers and consumers should recognize that Cloud services must fit and be managed within the consumers' IT environments. The IT landscapes of providers and consumers will be at various levels of adoption of the ITIL v3's (Information Technology Infrastructure Library) descriptive framework concepts and practices, at various maturity levels in their implementations of prescriptive ITSM (Information Technology Service Management) and BSM (Business Service Management) solutions, and at some level of readiness for Cloud Management solutions.

It is quite likely that each entity's service management solutions will have some simple, moderate or complex mix of ITIL, ITSM, BSM and/or Cloud services management, but it is very unlikely that any Cloud services provider or prospective consumer would be completely unaware of one or all of them.

Having said that, there is no absolute requirement from Cloud Computing for solution success that either one or both the Cloud service provider or consumer entities must be following ITIL practices, or have implemented ITSM and/or BSM systems. The provider could be using a system that provides sufficient service management tools and applications. The consumer could be just getting by with manual or desktop office applications to manage its raw IT resources. But it will hopefully become obvious after this chapter that ITIL practices implemented through ITSM and BSM systems are not only likely to be present, but may really power the best ways to ensure Cloud Computing success.

ITIL v3 clearly describes the optimal IT process to manage an enterprise's resources and activities, how IT management components operate and their interdependencies, and detailed descriptions for managing strategic and day-to-day activities in repeatable, process-oriented ways. It defines recommended Critical Success Factors (CSF) and Key Performance Indicators (KPI) to measure IT services and operations.

Control Objectives for Information and related Technology (COBIT) is another descriptive framework for IT practices that provides a collection of controls and Key Goal Indicators (KGI) to measure their success. ITIL and COBIT are descriptive frameworks that do not provide specifications on how to achieve their objectives, leaving this to interpretation and adaptation by practitioners.

ITIL and COBIT are perhaps best taken as a tandem. The question of COBIT versus ITIL does not recognize the synergism of these frameworks. The more mature IT organizations will apply both COBIT for control and ITIL for workflow to drive IT and business integration.

In addition to these descriptive frameworks, there are organizations that have produced or influenced IT management and operations standards, such as National Institute of Standards and Technology (NIST), the International Organization of Standardization (ISO), and others.

Finally, all of these sources provide IT service and business management recommendations, but the introduction of auditing and control requirements regulations such as Sarbanes-Oxley (SOX), the Health Insurance Portability and Accountability Act of 1996 (HIPAA), the Gramm-Leach-Bliley Act (GLBA), etc., have highlighted the regulatory importance of best practice implementation. Putting IT governance into action requires detail on what, when, how and who should accomplish the realization of IT strategic and tactical success.

Therefore, the descriptive frameworks of best IT service management practices must be implemented through the real business/IT processes provided by ITSM and BSM solutions.

This next section of the book develops a basic understanding of ITIL v3's IT service lifecycle management functions and processes to describe the best practices of IT service management, including the associated metering, analytics, and reporting activities. This limited introduction is provided here to foster the reader's recognition that IT support for the business is delivered as services, including IT support that is acquired as Cloud services.

A. ITIL's Foundation: Lifecycle Phases, Functions and Processes

ITIL v3 provides an integrated set of concepts and practices for Information Technology Service Management (ITSM), IT development, and IT operations management. ITIL is by far the most widely used framework of best IT practices. While most well-managed IT organizations also consider other sources, such as COBIT (Control Objectives for Information and related Technology) and ISO (International Organization for Standardization), the focus here is on ITIL given its dominance in ITSM practices.

The ITIL library is the result of various UK government agencies' (currently the OGC or Office of Government Commerce within the UK Treasury) work on providing "standard practices" for IT management.

ITIL v1 was built around a process-centric model of controlling and managing IT operations, becoming a collection of 30 best-practice-focused books with first publishing beginning in 1989.

ITIL, ITSM, and BSM Considerations

ITIL v2 began in 2000 to provide a consolidated collection of 8 logical sets of grouped process guidelines with greatest interest in the service management sets covering Service Support and Service Delivery.

ITIL v3 arrived in 2007 to provide the concept of a Service Lifecycle model consisting of 5 core phases with 4 primary functions and more than a score of defined processes that emphasize alignment of business drivers and IT resource allocations.

Figure 14-1 below is a graphical overview of the ITIL v3 Service Lifecycle Model.

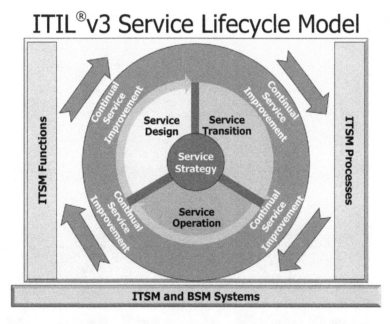

Figure 14-1 is an overview of the ITIL v3 Service Lifecycle Model.

The 5 phases of the service lifecycle should not be seen as separate or sequenced in a particular order, but more as a continuous cycle where each phase affects and is affected by the other phases in some key ways, e.g. the continual service improvement phase.

Figure 14-2 below is a more detailed overview of the Lifecycle provided by an exceptional graphic created by the ILX Group. It is exceptional both in its completeness and its presentation quality, representing that the ILX Group is a worthy training and consulting source for ITIL in the U.S. as well as the UK.

Note the identities of the five phases, the processes and functions of the phases, the cross-phase interconnections between the processes and contents, definitions of "databases" and "systems", etc.

This process model is introduced in greater detail, at a fairly high level, immediately after Figure 14-2, in the phase sequence of Strategy, Design, Transition, Operation and Continual Improvement.

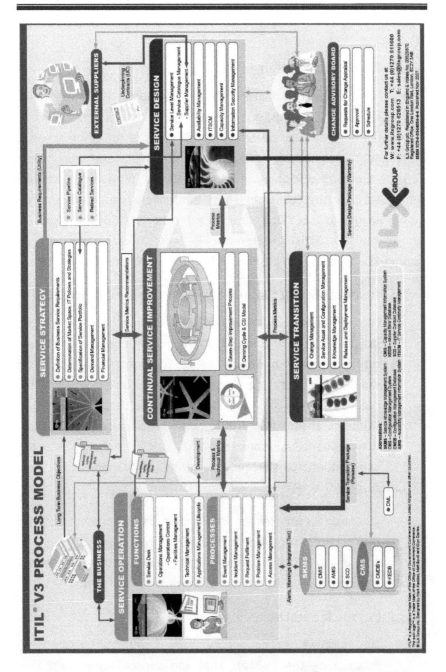

Figure 14-2 ITIL v3 Service Lifecycle Process Model from ILX Group.

1. The Service Strategy Phase

This phase is critical to the alignment of IT resources with the related business demands, with the goal of creating optimal value through the IT services delivered to the business. Strategy is the primary focus here, and that must begin with the business management clearly defining their strategic objectives and direction. This was discussed in some detail for the consumer entity earlier regarding <u>readiness for Cloud operations</u> (see "Consumer Cloud Readiness Considerations"). One of the most striking differences between ITIL v2 and ITIL v3 is around v3's emphasis on business-driven management of IT services.

a) Business Service Requirements Definition Process

ITIL v3 defines business services management as a best practice for the ongoing governing, monitoring, and reporting on IT and the business service it impacts. This practice emphasizes leveraging processes and technology to align the enterprise goals of IT and the business to become one. ITSM solutions ensure that ITIL framework processes and service management solutions converge for a holistic business contribution perspective in terms of costs, value, and competitiveness.

The business services requirements definition process looks at the services required to support business processes and provides definition in terms of each service's business value as a combination of its utility and its warranty to its users.

b) IT Policies and Strategies

The business service definition then is further refined by establishing the markets that the service serves and the corresponding market-specific refinement of IT policies and strategy.

c) Service Portfolio Management Process

A service portfolio should describe each IT service's attributes in business value terms. The value description follows a market-driven or user-centric perspective to provide requirements, value proposition, business case, priorities, risks, packaging and offers, cost and charges out (pricing).

Some service investments are required to just run the current business, while other service investments are necessary to grow the business, and others to transform into new business.

This process manages services across three possible states:

- Pipeline services that have been proposed or are currently being developed.

- Catalog services that are active and available for deployment, with service attributes such as description, functional specs, options, and prices.

- Retired services that have been decommissioned and therefore removed from the catalog.

d) The Financial Management Process

This process provides financial terms that describe the costs and business value of a service's required IT assets, and the associated operational management and support that is required. The key activities are around analyses for service funding and budgets, accounting for actual spend, and service chargebacks.

e) The Demand Management Process

This process enables the understanding and influencing of demand levels for a service through the analysis of business activity patterns and the user profiles for sources of demand. The process uses mechanisms such as premium chargebacks for certain types of service during high-load periods to influence demand patterns to fit optimal cost/benefit results for the organization.

2. The Service Design Phase

Service design activities occur whenever a new or changed service is required to meet changing business demands. This can be initiated from users, partners, etc., but the more significant changes would come from the Strategy Phase processes. The design activities should consider:

- Integration and interface fit within the overall service portfolio for management and control of services throughout their lifecycle.

- A service solution focus that provides requirements, performance specs, and all required resources and capabilities.

- The tools, technology and management architectures required to deliver the service.

- The implications to the processes required to design, transition, operate, and improve the service.

- The measurement systems, methods, and metrics for the service.

The design processes will result in the production of a service design package that contains business requirements, applicability, contacts, functional and non-functional requirements, service level requirements, design and topology (network and architecture), organizational readiness assessment, user acceptance test criteria, program outline, transition plan, operational plan, and service acceptance criteria.

a) The Service Level Management Process

This process ensures that the expected levels of service performance are completely defined and understood in agreed-on terms by all parties involved. Internally-supplied services are defined in OLA's or Operational Level Agreements. UC's or Underpinning Contracts set these criteria for services provided by external suppliers, such as external Cloud service providers. Agreements or contracts typically include:

- Service description

- Provider and consumer/user responsibilities

- Scope

- Service hours

- Service availability

- Service reliability

- User support expectations

- Contacts and escalations

- Service performance

- Security

- Costs and charges

The process encompasses monitoring and reporting actual service performance against the SLA's, performing reviews and creating plans to initiate service improvements, measuring user satisfaction, and participation in the negotiations of SLA's in general.

b) The Supplier Management Process

This process manages the external service providers (such as external Cloud service providers) and the services that they are providing to ensure the expected quality and value are attained as agreed. It covers supplier relationship management, negotiations and agreement in contracts with the supplier, aligning the UC's with the consumer's business needs, and maintaining the supplier policies and related Contracts data (in the ITIL-proposed Supplier Contracts Database). There is a number of possible co-, multi, or out-sourcing types of arrangements that define the expectations regarding the services.

c) The Service Catalog Management Process

This process is expected to create and maintain the contents of the service catalog to accurately reflect the services that are operational and available for deployment. The service catalog must be the single source of all information regarding the services from both business and technical perspectives.

The process ensures accurate maintenance of all interfaces and dependencies between each catalog service's details and:

- Other services in the catalog, including any services that are supporting this service.

- The related service portfolio contents.

- Configuration Items (CI's) within the service catalog and the CMS (Configuration Management System).

A service's hierarchy may consist of:

- Business services

- Supporting services

  - Infrastructure server, network and data management services

  - Platform and application services

- Generic, shared, commodity services

- Externally-provided services

The end-consumer of the services should be clear despite the potential for many indirect layers, likely present with external Cloud service providers or virtualized resources in general.

d) The Capacity Management Process

This process ensures that IT capacity is mapped to current and agreed business-driven needs that have been cost-justified. It works with financial management and demand management processes to maintain the desired balance between resource supply and demand.

The primary capacity management activities are:

- Capacity modeling to forecast what-if supply and demand scenarios.

- Application sizing for requirements associated with new or changed business processes.

- Capacity planning to forecast capacity increases and decreases.

- Reactive implementation of strategies on short-term basis from Service Strategy demand management analyses.

- Tuning modifications from the Change Management process that are required to improve IT resource utilization.

- Performance monitoring and reporting, measuring and tuning of services and related components to improve capacity utilization levels.

- Storage of all business, service, and component capacity data in a Capacity Management Information System database.

These activities are performed within the contexts of 3 capacity management sub-processes:

- Business capacity management driven by changes in business requirements.

- Service capacity management driven by service level agreement expectations

- Component capacity management driven by optimal performance assurance of resources.

e) Availability Management Process

This process ensures that the level of service availability delivered by services is driven by the demands of the business in a cost-effective manner. It can do this through reactive activities that are initiating steps based on unavailability incidents and remedial actions taken. It could also employ risk assessments that produce prediction-based actions to resolve anticipated issues to avoid unavailability incidents. ITIL defines an Availability Management Information System (AMIS) that provides this level of monitoring, analyses, and assessment.

In addition to the measurements and considerations (see "Availability, Reliability, Maintainability and Metering") provided above, a serviceability measurement indicates the ability of an external (e.g. Cloud) service provider to meet the terms of their contract regarding service levels, etc.

f) IT Service Continuity Management Process

This process (ITSCM) ensures that the IT resources and their provisioning can be recovered at the desired levels within agreed timelines after a disaster or other continuity disruption. The process encompasses planning for disasters, implementing processes and mechanism to support policies and expectations, and then as-needed managing the recovery to business continuity. Deliverables include:

- ITSC plans and recovery plans.

- Business impact analyses.

- Risk analyses.

- Continuity and recovery mechanisms implemented and proactively enhanced.

- Assessments of change impacts on ITSCM plans and procedures.

- Assure supplier (e.g. Cloud service provider) contracts support expected ITSCM practices.

g) Information Security Management Process

This process aligns IT and business security objectives, policies, mechanisms, analysis and reporting attributes. These attributes (see "Cloud Security Considerations") ensure confidentiality, integrity, availability, and trust associated with the consumer's assets, information, data, and IT services:

- Availability is assured by making information available and usable as required by the business practices, and by enabling all solutions to resist attacks, prevent failures, and recover from failures as quickly and elegantly as possible.

- Confidentiality is assured by limiting access to those with rights and privileges.

- Integrity is assured by maintaining accurate, complete, and uncorrupted information protected from unauthorized modifications.

- Trust in business transactions and exchanges with business partners is assured by authenticity and non-repudiation practices.

ITIL suggests that an Information Security Management System (ISMS) should contain standards, procedures and guidelines supporting security policies. This system would be expected to support the activities of this process across the entire service lifecycle:

- Security policy development and maintenance.

- Communication, implementation and enforcement of security policies.

- Scheduling and execution of security audits, penetration test, and reviews.

- Implementation and continual review of security controls.

- Monitoring and management of security-related incidents.

- Analyses, reporting and reduction of frequency and impact of security breaches and incidents.

h) Risk Management Process

Risk is uncertainty about an outcome, such as technological risks, commerce or financial risks, information security risks, etc. The consumer should manage its risks in a visible, repeatable, and consistent manner across the entire service lifecycle. Some risks are managed through business impact and what-if analyses, while others are recognized by looking for threats or vulnerabilities.

This is a sub-process that makes use of a risk framework to employ a series of well-defined steps to assure better decision-making on a proactive basis. It accomplishes this through a good recognition, identification, and understanding of risks and their associated impacts on business practices, processes, and results. The process ensures identification, analysis and response to risks that might adversely affect realization of business and/or IT objectives. The response to risks depends on impact, and involves avoiding, accepting, controlling, or transferring them to some other party.

i) Compliance Management Process

This also is a sub-process, and it assures design conformance with service level, contract, or other expected performance requirements. It is achieved through the identification of the applicable requirements (defined for example in laws, regulations, contracts, strategies and policies), an assessment of the state of compliance, and an assessment of the risks and potential costs of non-compliance against the projected expenses to achieve compliance.

The result may be to re-prioritize, fund and initiate any corrective actions deemed necessary. The process encompasses all aspects of defining, planning, implementing and scheduling compliance reviews and audits, as well as monitoring and reporting on compliance.

j) IT Architecture Management Process

This is yet another sub-process, and it defines a blueprint for the technological landscape that considers changes to service strategy and/or designing new services with newly available or expected disruptive technologies.

ITIL v3 provides guidance on IT architecture issues in a "technology-related activities" chapter. A well-defined architecture blueprint is critical for IT service management to be proactive when possible or responsive in a timely and effective manner. A poor architecture will have ITSM chasing gaping IT service management (and business performance) wounds with band-aids.

The process should readily support a request to change or extend the IT Architecture, usually issued within the Service Design process when the introduction or modification of a service is not possible within the constraints of the existing application, platform, infrastructure and information / data architectures.

3. The Service Transition Phase

New and modified services are transitioned from the Service Design Phase into the Service Operation Phase through the Service Transition Phase, hence the emphasis on transition. Transitioning services must meet requirements and not affect business processes, other resources or services in any undesired ways. Quality releases are built, configured, tested, and deployed without business disruptions.

a) Knowledge Management Process

This process ensures that meaningful data and information is made accessible throughout the IT service's lifecycle, reducing the need to rediscover knowledge and thereby improving the efficiency wherever data is critical. The process deliverables can be understood in the DIKW (data, information, knowledge, and wisdom) knowledge lifecycle construct.

ITIL describes a Configuration Management System (CMS) manages all CI's or Configuration Items (IT components – IT Services, hardware, software, buildings, people, and formal documentation – that need to be managed to deliver an IT Service). Within the CMS, ITIL calls for a Configuration Management Data Base (CMDB), containing information regarding incidents, problems, known errors (the known error data base, KEDB), changes, releases, employees, suppliers, locations, business units, service consumers and users, etc. It also provides the tools to use the CMDB's data, including collecting, storing, managing, updating, and presenting.

ITIL also describes a Service Knowledge Management System (SKMS) that encompasses the CMS and adds the remaining service management data and tools associated with staff experience levels, peripherals, supplier and partner requirements and abilities, user profiles, etc. The SKMS stores, manages, updates and presents the information the provider needs to manage data and information across all processes within the overall service lifecycle.

b) Service Asset and Configuration Management Process

This process maintains the CMS, and should include automation of items' discovery, inventory, audit, network management, and other interfaces to the CMS.

The service assets could be any capability or resource that can contribute to the delivery of a IT service, including management, organization, process, knowledge, people, information, applications, infrastructure, and financial capital.

The process identifies configuration items (CI's), plans CI management, controls CI integrity, provides status accounting, CMDB audits and reviews, stores, and provides information about the CI's, the relationships between CI's, and the service assets inventory throughout the service lifecycle.

c) Change Management Process

This process employs standard methods and procedures to control any and all alterations of CI's, including the addition, modification or removal of hardware, software, application, environment, systems, desktop builds, documentation, etc.

It is the primary contributor to the integrity of the CMDB through its RFC (request for change) control, assessment, authorization, planning, coordination, review and close activities. The RFC's may come from anywhere, across a range of requests involving a standard pre-approved change; a normal change that follows the normal expected assessment and authorization by a Change Manager or a Change Advisory

Board; or an emergency change that uses a fast response process for Emergency Change Advisory Board approval, etc.

d) Release and Deployment Management Process

This process places services into active operation by deploying new release packages into production, and transitioning support from development and test to operation. The activities of this process include:

- Release Policy and Charter definition.

- Release planning.

- Build, test, and deployment planning and preparation.

- Build and Test execution.

- Service tests execution.

- Deployment and retirement execution.

- Verify deployment.

- Initial post-deployment support.

- Review and close of deployment.

- Review and close overall service transition.

- ITIL also calls for activities within this process that maintain the Digital Media Library's (DML) contents regarding software licenses, source code, etc., and the Definitive Spares (DS) (hardware) storage for IT assemblies, etc.

4. Service Operation Phase – Functions

This Service Lifecycle Phase provides management of the IT services through the coordination, delivery, and managed execution of IT services provided to business consumers and users. There are 4 major functions that make up the operations roles and automated measures employed within Service Operations.

a) Service Desk Operations Function

This function is one of the most visible aspects of IT service management. It supports service provisioning at the agreed levels of performance, and acts as the single point of contact for all service incidents and requests to assure the restoration of normal service operations with the least disruptive impact to business.

This is much more than a help desk or call center function, as it owns the reported incident process from start to resolution and closure. The desk may be a local, central or regional physical structure, or perhaps a virtual desk with incidents handled by agents distributed geographically. Extensive use of self-service, known errors handling, and proactive event management practices are likely to reduce the burden on service agents.

b) Technical Management Operations Function

This function's role supports the business processes through the planning, implementation and maintenance of a stable technical infrastructure. It provides well-formed, resilient, cost-effective architectures, and applies the proper diagnosis and resolution capability to restore service. It draws its best staff resources from both Design and Operations people.

c) IT Operations Management Operations Function

The role of this function is to use repeatable and consistent actions to perform day-to-day service operation assurance activities, including monitoring and control, dashboard and console management, job scheduling, backups and restores, print and output management, and provides all aspects of Facilities Management, whether on-premise or outsourced.

d) Application Management Operations Function

This function's role is to develop, maintain, and support well-designed business applications that effectively interface where needed with other platforms or applications. It ensures that the applications' functional, technical, and performance requirements are defined, that these requirements are met in making either the build or buy applications decision-making processes, and provides ongoing support and enhancements for production.

This should be a critical interface function for legacy IT applications hosted on public Clouds, or for the externally-provided platform or application Cloud services that must be included in the consumer's ITSM and BSM systems' scope.

5. Service Operation Phase – Processes

The previous discussion covered Operation Phase Functions, and this discussion focuses on the Operation Phase Processes.

As stated above under functions, the Service Lifecycle Phase provides IT service management through the coordination, delivery, and managed execution of IT services provided to business consumers and users. There are 5 major processes that support this phase, where Incident Management, Request Fulfillment and Access Management processes are typically carried out by the Service Desk function. Event

Management and Problem Management processes are primarily performed by the other Operations functions.

a) Event Management Process

An event is defined as a change of state having significance for the management of a configuration item (including IT services). This process delivers operation monitoring and control through capabilities around detection of events, filtering or rationalizing them, determining and executing the desired response action for the event.

Some actions may provide informational updates while others provide alerts that something requires attention, and other events trigger automated actions (without human intervention) performed immediately. Event management is employed to detect, communicate, and/or execute actions that automate backups, batch processes, dynamic load balancing or service optimization, etc.

b) Incident Management Process

This process restores normal agreed-on service levels of performance to service operations as quickly as possible while minimizing the adverse impact on business operations, assuring best quality and expected availability levels of services. It employs predefined incident models that provide a chronological procedure of steps encompassing all dependencies or co-processes, assigned responsibility roles with action schedules, scales, thresholds, escalation, and evidence-preservation, etc.

Standard incidents may have automated handling, escalation, and management of the incident management process.

c) Problem Management Process

This process manages the lifecycle of problems. Problems are incidents that cannot be resolved through error knowledge and resolutions knowledge previously gained and stored, but must instead be resolved through the diagnosis of the root cause and the determination of a resolution to the problem. It ensures the implementation of the problem resolution through interfaces with change, release and deployment processes.

Proactive problem management uses analysis of incidents for patterns to identify trends, predict problem occurrence, or isolate significant problems candidates.

d) Request Fulfillment Management Process

This process uses preapproved, consistent and repeatable standard services to fulfill standard requests from end-users where the issue is not considered to be business process-disruptive. These requests are usually fulfilled by automated, self-service processes. One example would be preapproved downloads of desktop applications by an employee.

e) Access Management Process

This process provides capabilities around granting or declining of rights and privileges for authorized access to use a service, through activities including user verification, priority with resource availability consideration, etc. It interfaces with processes in the Design Phase regarding information security and availability management, and with processes in the Transition Phase regarding change and configuration management.

6. Continual Service Improvement (CSI) Phase

The Continual Improvement Phase ensures ongoing improvement of all of the ITIL Service Lifecycle Phases, their functions and processes. This phase effectively ties all of the elements in the Lifecycle together through comprehensive metrics covering the capabilities and the services they provide.

Figure 14-3 below is a representation of the Deming PDCA improvement cycle put forward by ITIL.

The Deming Cycle

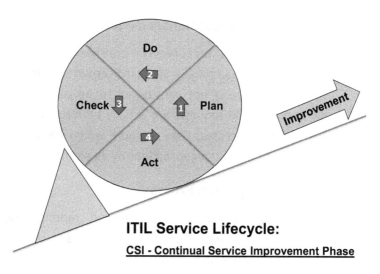

ITIL Service Lifecycle:

CSI - Continual Service Improvement Phase

Figure 14-3 reflects the ITIL CSI improvement PDCA cycle.

a) Seven-step Improvement Process

This process utilizes a framework structure for defining, analyzing and using metrics to improve services, the functions and various service management processes. It requires the definition of service vision and strategy within the Strategy Phase, and the definition of service tactical and operational goals within the Design Phase.

With these definitions in place, the seven process steps are:

- Define what should be measured.

- Define what you can measure, and do an analysis of any gaps between current abilities and desired measurement.

- Gather the data.

- Process the data.

- Analyze the information formed from the data.

- Present and use the information.

- Implement improvements.

b) Service Measurement and Reporting Management Process

This process coordinates the design of technology, process, and service metrics with the collection of data and reporting of metrics from all functions and processes. The CSI phase applies the seven-step process through relationships and exchanges with each of the other phases:

- Service strategy in general and specifically the Service Portfolio drive the service vision.

- Baselines taken from Service Portfolios, Service Level Management, and Financial Management for IT processes establish the current service performance position.

- Strategic targets for service performance are derived from the Service Portfolio and Service Measurement and Reporting processes.

- CSI and all ITIL functions and processes drive the determination of paths to improvement.

- Service Measurement and Reporting process provides the analyses to determine if expected improvements were achieved.

- The CSI phase's activities provide the overall ongoing drive towards continual improvement.

B. ITSM and BSM Solutions

The ITIL v3's best practice descriptions are implemented in prescriptive ITSM and BSM solutions.

1. IT Management Solutions

The service provider's ability to deliver effective Cloud management can be directly related to its ability to effectively manage its Cloud services within an integrated, modularized, standardized, and automated service management context. But it is the consumer's ability to manage acquired (e.g. Cloud) IT services throughout the Cloud service lifecycle that may be of even greater importance to the success of widespread adoption of Cloud Computing solutions.

ITSM solutions are typically drawn down into large volumes of control and data of the day-to-day operations. (Note that the functions addressed in the ITIL v3 Service Lifecycle exist within the Operations Phase.) Though driven by ITIL v3 practices, ITSM solution deliverables may still predominantly focus on IT infrastructure and the technology.

ITIL, ITSM, and BSM Considerations

Many informed predictions indicate that the positive business results associated with Cloud Computing solutions will drive significant increases in Cloud Computing adoption over the next few years. Similar positive benefit-driven predictions exist for BSM adoption.

BSM solutions define, implement, maintain, monitor and measure services from a business perspective using a set of management-enabling software tools, processes and methods. They enable IT operations to be focused on business-valued priorities and services rather than the details provided for the individual configuration item, etc.

BSM brings timeliness and business-relevancy into IT management decisions. BSM solutions go beyond the alignment of IT and business objectives. They move IT Management beyond managing IT operations and IT service lifecycles toward the integration of business goals, objectives, priorities, values, and processes with IT management of all IT resources, in strategies, plans, and practice. The same level of focus that ITSM may place on the servers, storage, network and facilities should be expanded by BSM to encompass business services, end-to-end business processes, activities, applications, transactions, etc.

A BSM platform's applications will produce reductions in IT costs, consistently positive business impacts, improved service quality levels, proactive use of "good" risk and removal of "bad" risks, with much greater transparency. It produces greater modularization and simplification along business driver vectors, while standardization and automation of IT processes deliver efficiency and effective decision support.

BSM solutions tend to be found more frequently in larger enterprises, and the integration of BSM and Cloud Computing is more likely to be an area of strategic importance in those companies where a productive BSM solution is more mature. It is not surprising, then, that a major challenge to Cloud adoption might come from enterprises whose significant investment in ITSM and BSM leads them to fear losing control over their applications in the Cloud. The reality is that a productive BSM solution should be recognized as a critical enabler for businesses that are placing their business services into a Cloud Computing solution, and that Cloud Computing adoption should in turn become a major catalyst for a much broader base of BSM adoption.

2. ITIL v3 and BSM Solutions

From a Cloud services perspective, it is important to note that there several specific key principles, concepts, and practices that have been introduced in ITIL v3 which BSM solutions are intended to directly support.

- **The integration (beyond mere alignment) of IT and business.** ITIL v3 establishes Business Service Management as a best practice of governing, monitoring, and reporting on IT and each affected business service. There is a new singular perspective that understands how business affects IT and how IT affects business, yet is capable of communicating in the context of its audience, business terms or IT terms. BSM solutions leverage ITIL practices and technology to integrate IT and business as one set of goals focused on delivering business value and ensuring that IT management is a primary enabler of strategic and tactical business success.

- **A comprehensive service lifecycle management approach.** Every phase in the service lifecycle is focused on producing business value. BSM solutions provide end-to-end, closed-loop process management to achieve optimal business results supporting the service (SLA's) and operational level agreements (OLA's), underpinning contracts (UC's), policies, rules and thresholds for

15

Initiatives for Integrated Service Management

Brainstorming discussion

Discussion Areas in This Chapter

- *Integrated process and systems*

- *Solution outline*

- *Cloud commitments*

Initiatives for Integrated Service Management

This chapter is focused on providing a "brain-storming" discussion on how thought leaders in the area of ITSM, BSM and Cloud management solutions might pursue comprehensive integrated solutions that serve the shared interests of Cloud service providers, consumers, and enablers.

A. Integrated Service Management Processes and Systems

Cloud management and BSM solution vendors appear to be in the early stages of integrating BSM and Cloud management applications, tools, processes and systems. The promise of comprehensive information and application integration is a very complex challenge when considering the concept of integration across multiple Clouds with multiple providers and multiple consumers.

1. BSM and Cloud Management

In this context, the optimal solution architecture might support a singular solution paradigm that serves providers and consumers. The design objectives of tools and applications should primarily focus on enabling both Cloud service providers and consumers to have a real-time, comprehensive, holistic and seamless view of their shared and respective independent business services across one or more Clouds and any related non-Cloud internal data center-based applications.

The new Cloud Management "business service management" solution's scope would support an environment where the provider works with each consumer through every phase and process of the Cloud service lifecycle to produce the desired integration of:

Initiatives for Integrated Service Management

- The provider's Cloud services, business services, internal IT services, and supporting external (sub-) provider services (partners, etc.) that enable their end-to-end business process. Each service should be understood and defined in the context of its impact on the provider's Cloud and business services, and the associated business/IT components, attributes, challenges, goals and objectives.

- The consumer's business services, internal IT services, etc. mapped to the Cloud services, to produce the end-to-end business processes. Each service should be understood and defined in the context of its impact on the consumer's business, and the associated business/IT components, attributes, challenges, goals and objectives.

- The "intersection" elements of the above two perspectives, maintained and offered from one master integration Cloud Management and BSM construct, that enables the above sets of unrelated and related processes, functions, information, and solution components, within a verticalized cross-business shared-context perspective.

The current scope of some solid BSM solutions already comes close to supporting the first bullet above, delivering the service provider's management of their resources through numerous tools for diverse types of resources, including servers, appliances, virtualization, storage, network, security, load balancers, provisioning and automation, application software platform monitoring, application software monitoring, etc. capabilities within a Cloud. The provider's BSM solution would still deliver "integrated business/IT service" support but should be expanded as needed to manage the multi-consumer (and their multi-business impact implications) environment brought about by virtualized infrastructure and multi-tenant software.

The BSM solution for the Cloud service consumer is expressed in the second bullet above. This BSM solution would require its various solution process, object and data models be expanded to support the additional views that would establish the provider's Cloud as another set of integrated business services to be managed within the consumer's standard BSM processes. Yes, the service allocations and consumption are abstracted from the physical resources, but there is sufficient data that can be defined to logically associate a virtual resource tethered to a physical resource during the service deployment. It becomes a matter of design and realization.

The architecture and design of the overall solution should allow the providing and consuming parties of a Cloud service to choose their initial, mid-term, and long-term "balance" of sharing across BSM functions, processes, systems and databases. Cloud Management should be able to expose the management of their resources at the most abstract and granular views, and everything in between, based on the preferences agreed on by provider and consumer. The level of integration across the provider and consumer views could be a "configurable" added value-priced component, or a standard component within the Cloud service standard package offered as a competitive differentiator at no additional charge.

Remember, the business challenge is not about IT, it is about the integration of IT services and business services to support end-to-end business processes, consistently producing the competencies and capabilities that enable the business to be successful. Business processes that are delivered in part or completely through business services provided by external sources should not be excluded from the scope of business service management simply because Cloud services are more complex for providers and/or consumers to manage.

Figure 15-1 represents a very high-level representation of the 5 fundamental initiatives that thought-leading IT and Cloud management vendors might pursue, with emphasis on integration and automation, first within the service provider's BSM and Cloud Management systems, and then with the Service Consumer's BSM applications. If an integrated service management solution is to be taken seriously, it should be playing aggressively with strategies (and execution) across all 5 of these initiatives.

Cloud Initiatives for Management Solutions

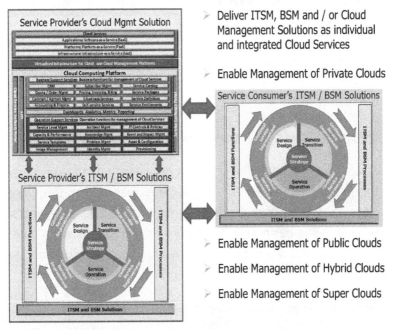

> Deliver ITSM, BSM and / or Cloud Management Solutions as individual and integrated Cloud Services

> Enable Management of Private Clouds

> Enable Management of Public Clouds

> Enable Management of Hybrid Clouds

> Enable Management of Super Clouds

Figure 15-1 identifies 5 initiatives for integrated service management

- **Deliver Cloud Management, ITSM and BSM solutions as individual and integrated Cloud services.**

   Service management is a common challenge that is shared by each of these three solution areas. As business opportunities push increases in Cloud adoption, it will become clearer where the business-enabling integration challenges of the Cloud providers and consumers lie. As this happens, the focus moves from IT technology management, to IT Services Management, to pre-Cloud business-integrated IT Management, to Cloud Management, and finally to an Integrated Cloud Services Lifecycle Management. On-demand solution deliveries of these management systems would also enable the solution vendors to reach new markets while sharpening the focus on what a provider really needs to manage in the various Cloud structures. A comprehensive on-demand solution that integrates these management solutions would soon become very popular in Cloud adoption circles, generate steady streams of substantial revenue for the service providers, and fully realize the promise of ITIL v3 in a Cloud Computing era.

- **Enable Provider's Management of Private Cloud Services**

This is where it all begins for most providers and consumers, and the predominantly infrastructure focus means that the biggest players will likely come from the ranks of existing IaaS Providers or their enabling partners. Integrated management must be able to work here best if it is to have a chance in public, hybrid or super Clouds.

- **Enable Provider's Management of Public Cloud Services**

Public Clouds are where the greatest amount of broad and deep Cloud innovations is likely to occur. And one of the greatest challenges for public Cloud applications, platforms, and infrastructures will be integration. As long as public Cloud service providers are seen as individual silos of solutions, Cloud adoption will be slowed. So the advent of integrated service management supporting business-driven IT management is a very good thing. If the integrated service management vendor can successfully offer on-demand services in a public Cloud context, it should know how to enable others to do likewise. If that vendor has already provided integrated service management that enables the Cloud management of a private Cloud, they also know how to manage this market segment's Clouds.

- **Enable Provider's Management of Hybrid Cloud Services**

  The natural extension from a private Cloud to adoption of one or more public Clouds is directly through a focus on infrastructure. The integrated service management solution must be able to deal with this market segment, especially if it can cover the scope of at least 2 of the 3 initiatives above.

- **Enable Provider's Management of Super Cloud Services**

  The ultimate integrated service management solution is one Cloud that encompasses a combined network of multiple interoperating Clouds in support of an enterprise and its business partners. While the hybrid Cloud extends the infrastructure resource availability to provide elastic scalability, the Super Cloud model involves the integration of applications supported by two or more Clouds into related or unrelated end-to-end business processes. Again, if the previous 4 initiatives are supported by an integrated service management solution, it is quite possible that this solution can also serve the demands of this initiative.

2. An outline of an integrated management solution

This is a high-level representation of the broader integrated service management solution within the expanded scope and capabilities as generally defined above. It is important to recognize that a comprehensive ITSM – BSM – Cloud Management solution would serve the needs of all parties: those needs uniquely associated with the provider, the needs uniquely associated with the consumer, and the needs commonly applied in their shared collaborative efforts.

The comments below should be applied to all of these three views (provider, consumer, or shared) unless explicitly assigned to one or two of these three views.

a) Strategy

There are several strategic research, analysis (see "Consumer Cloud Readiness Considerations"), and decision-making business processes that drive an enterprise's strategic planning (see "Use cases for software provider's internal operations ") for products and operations. The enterprise should perform a dedicated strategic planning function that identifies business opportunities, competencies, differentiation options, and defines trends in markets, and technologies (especially potentially disruptive technologies). These semi-annual or annual activities would typically result in isolating two or three major initiatives worthy of further consideration, ultimately leading to formal development and consideration of their related business cases. Approved changes in business strategies could lead to product extensions in existing markets, new platforms and products, expansion into new markets, dropping products or product lines, etc.

All of the BSM service strategy processes are engaged to analyze the business alternatives and options, the capital or non-recurring (capex) expenses, the recurring operating expenses (opex), the expected changes in revenue, the upside and downside risks, etc.

The consumer's adoption of a Cloud Computing solution should begin with the definition of business case alternatives supporting "buying" services from an external provider rather than continuing to make or build these capabilities in-house. The initiative is translated into business process requirements and the IT services that are required to support them. This involves:

- The definition of business requirements for the expected business model and business processes that will be supported. The Cloud services that are to be procured should be sufficiently detailed to drive the definition of design, transition and operations support elements for contract terms, systems implementation and integration, etc. These terms establish the preliminary utility of the expected Cloud service assets, and identify the high-level interdependencies between all affected services, etc.

- The determination of demand management issues associated with demand patterns and user profile expectations for the business services. These Cloud service demand details will drive discussions regarding pricing, service availability, and capacity, scalability, and elasticity levels.

- The definition of the IT policies and strategies that will be applied to the business processes. They include the expected protocols between consumer and Cloud provider that will be required to ensure that the business services are operating at the desired levels.

- The integrated business/IT financial data, provided by the IT financial management sources, that is used in the business case to justify the decision to move forward with the implementation of the Cloud services initiative.

- The service portfolio contents that describe the business value associated with implementing the new Cloud business service. If the Cloud services will replace an existing set of business services, these would also be identified to recognize end-of-life and service transitions challenges.

- A set of high-level Cloud service metrics should be defined and passed on to the CSI and Design processes. It is critical that the proper metering, analytics, and reporting expectations are met, especially as they determine the integrity of business case arguments.

b) Design

The scope, behavior and business service requirements that must be met have been defined. The design for the Cloud services now requires that the architectural issues around tools, technology, performance, measurement systems and methods, metrics, etc. be addressed. The design should consider the implications to all affected processes within the design, transition, operations, and service improvement phases.

All of the design phase processes are involved in the specifications associated with the expected solution. They will collectively produce a service design package that establishes the expectations between the provider and consumer, and the roadmap for the transition and operations phases. The package contains the pertinent attributes of the SLA's, OLA's, UC's; the business requirements, applicability, reciprocal provider and consumer contacts, functional and non-functional requirements, service level requirements; the initiative program, solution architectural design and topology, enterprise and organizational readiness assessment, user acceptance test criteria, transition plan, operational plan, and service acceptance criteria.

- Service catalog management will provide the comprehensive business and technical attributes of the Cloud services, their interfaces and dependencies, in a hierarchy that defines any associated business services, application services, platform services, and infrastructure services corresponding to the service assets and other CI's.

- Service level management is critical to this process. The expected performance attributes and levels for the Cloud services must be defined in clear terms for monitoring, measuring, resiliency, maintainability, availability, accessibility, portability, costs and charges, etc. The API elements and mechanisms provided for sharing the support activities between the provider and consumer should be identified and committed, and include reviews and planning for creating service improvements, defining and measuring performance and satisfaction levels.

- IT service continuity management will define how the consumer's data in the provider's Cloud and the associated Cloud services will be recovered at desired levels and within agreed timelines. A shared provider-consumer plan for recovery and continuity should be defined, and the mechanisms implemented and enhanced as needed.

- Supplier management is also critical to Cloud services adoption and ongoing success. It starts with managing the provider-consumer relationship, includes the UC contract negotiations and data integrity in the Supplier Contracts

Database (SCD), and alignment of UC contents with consumer business services and with the internal OLA's supporting these services.

There should be considerable control and audit integration without undue redundancy between the service assets in the CMS (perhaps in the CMDB itself) and the Cloud services ascribed to in the contracts' and agreements' logical "artifacts" within the SCD. It is also clear that there must be some metadata level that maps the provider's virtual resource allocation to an ID that can be used in all related cases by the consumer.

- Capacity management will focus on;

  o Sizing the requirements for the business services, to determine the impact of the changes in business service capacities driven by the capabilities of the new Cloud services. This will be pooled in capacity modeling forecasts with any offsets produced by the reduction in capacity demands from any replaced legacy business services.

  o Service capacity analyses are driven by the various service level agreement parameter options being considered. Scalability and elasticity ranges should be addressed.

  o Capacities should be addressed at all component levels across the application, platform and infrastructure layers where control and risk-of-failure opportunities exist.

  o There should be sufficient capacity details to address the objects and elements that need to be exposed in the API's provided by the provider. Service performance tuning, monitoring, measuring and reporting expectations should be committed in the underpinning contracts with the provider, and in the related operation level agreements for making the Cloud service operate seamlessly with the business service users, while assuring adequate provider and consumer IT ownership of the services.

o The capacity data for the Cloud services should be available in the Capacity Management Information System (CMIS) and support on-demand capacity planning activities that enable shared provider-consumer forecasting of potential changes in capacity.

• Availability management should determine the balance of supply and demand for the Cloud service considering the business service asset value (utility, warranty) and the tipping points associated with service level expectations. The Availability Management Information System (AMIS) should support the level of data required for shared provider-consumer monitoring, analysis, assessment, and management activities.

• The information security management activities should align the shared provider-consumer security objectives, policies, mechanisms, data, service dependencies, etc. The Information Security Management System (ISMS) should accurately reflect the agreed security policies, guidelines, standards, and procedures that assure expected levels of availability, confidentiality, integrity, trust authentication and non-repudiation practices.

There should be clear expectations and shared provider-consumer responsibilities regarding enforcement, control, access, implementation, reviews, audits, penetration tests, monitoring and management of security incidents, security improvement activities, etc.

- Risk management should be employed in the initial and ongoing assessment of business service impacts from the Cloud services. The consumer and provider should collaborate on defining potential risks and business impact analyses, and how to best apply the risk assessment and management framework.

- Compliance management is applied to ensure conformance by the provider and consumer with the contract terms regarding policies, guidelines, agreed service and performance levels. Both parties should be involved in varying degrees with defining, planning, and execution of reviews, audits, and reporting on compliance.

- IT architecture management should ensure that the overall solution design of the Cloud services interoperating with the other business services is well conceived and properly implemented. The consumer should have a complete picture of the Cloud service's model, its components and their business services impacts.

c) Transition

The Cloud provider's services must be transitioned from a collaborative design into the productive operations objects and processes. The Service Design Package contents delivered to the service transition phase will define the business service warranty elements. The various transition processes and activities will produce a Transition Package or Service Release for deployment into operations.

All of the strategy and design components regarding provider-consumer shared and collaborative initialization and service maintenance must be transformed into solution production elements to ensure the expected operational outcomes for the service. The standard Cloud service attributes must be customized if needed, with a stable build, configuration, test, and deployment requiring the synchronized efforts of provider and consumer.

- Knowledge management is at the core of successful Cloud services. The Service Management Knowledge System (SKMS) is at the center of the integration, control, and orchestration of the service management processes throughout the initialization, implementation and production assurance processes. All forms of knowledge exchange and sharing between the provider and the consumer should be reflected in the CMS's CMDB and KEDB, etc.

- Service asset and configuration management will assure the integrity of the CMS contents and the effective use of the Cloud service API's and service delivery API's, to provide integration of consumer databases and tools with their provider counterparts.

- Change management will enable the control of addition, modification and/or removal, as needed, of all CI's associated within the provider's Cloud service, from initialization through any ongoing change requests. It will ensure the ongoing integrity of the CMDB through its request control activities. Again, the change in any facet of the SKMS, and especially in the areas around the CMS's CMDB, should be tightly integrated for both the service provider and consumer counterparts.

- Release and deployment management places the Cloud service and its associated business service assets into active operations when the transition package is deployed as a Release. The close (early) support teams of both the provider and the consumer will use the implementation period to quickly get to production status and work out the kinks of their collaboration process expectations.

d) Operation Functions

The operations functions will work with the transition team as it takes the Cloud service's release package into production, and then manage the coordination, delivery, and execution of the Cloud service. The consumer's operations functions must work in close cooperation with the Cloud service provider's counterparts.

- Service desk collaboration is critical here, because the single point of contact between the consumer and the provider must be clearly defined and supported without fail. The best solution depends on several factors, but the key objective is that the consumer's service level expectations are met, and that both the provider and the consumer are completely on the same page. Automation and self-service should resolve most low-end issues, and both service desk teams should be equally aware of these requests and incidents. Incidents and problems should follow a prescribed path of resolution that recognizes the consumer's closeness to their processes and fit as the single point of contact, and the provider's better awareness of the possible solutions for the root cause of the problems.

- The technical management operations will draw upon staff from both the consumer and the Cloud service provider through planning, implementation and maintenance that assures a stable technical architecture. As the consumer's design and operations teams become more proficient in the diagnosis and resolution capabilities, the provider's support team becomes the escalation team that works on the most challenging problems, especially those requiring changes to the service provider's service resources.

- IT operations management will be predominantly a Cloud service provider function with a consumer "governance" monitor. As long as business service levels are met as expected, this function is sharing the information about the provider's day-to-day operations at the agreed-on levels in a timely and informative manner with the consumer.

- Application management is a critical function in all Cloud services, as business services are being consumed through applications that are supporting business processes. It will be a critical factor in the definition of the application's service requirements, its design, its effective implementation, and its support and enhancement as needed. For Cloud-provided application software platforms and application software, the provider and consumer should be creating and managing the effectiveness of the software applications through this function, providing a very critical role in working through shared support and development teams' issues.

e) Operation processes

The operations processes span both proactive and reactive activities that are focused on keeping services up and operating at the agreed levels of performance to support stable normal business operations.

- Event management is primarily performed by technical, IT operations, and applications management functions. It requires that the provider and consumer agree on the definition of events, the response for these events, on the related automation triggers and their actions, and the sharing of the responsibilities in managing the monitoring and responses.

- Incident management is predominantly performed by the service desk function of the consumer, where incidents are likely distributed or allocated between consumer and provider based on known errors and resolutions versus new problems, as well as the complexity of the issues as standard issues versus complex problems best resolved by the provider's support and development functions. But the consumer's service desk should retain its position as the consumer's single point of contact if it wishes to remain the primary enabler of quick returns to normal service levels and least negative impacts for the disruptions to the Cloud service and the business processes that it enables.

- Problem management is also a process initiated through the consumer's service desk, but the deep analyses and resolution is predominantly performed through the work of the provider's technical and applications management teams. If the consumer's service experiences no disruptions to the agreed service levels, this is entirely recognized (proactively perhaps) and handled within the provider's problem management team.

- Request fulfillment management employs preapproved, consistent and repeatable standard services to fulfill standard requests of the consumer's users. These services are usually automated self-service processes, predominantly using the consumer's service desk as an automated conveyance of service issues but are also often supported directly by self-service from the service provider.

- Access management protocols and mechanisms are initially defined collaboratively by the Cloud service provider and consumer teams, especially the design teams' information security and availability processes, and the transition teams' change and configuration management processes. But the ongoing management of this access management process should be performed by the consumer's service desk function.

f) Continual improvement

The ongoing improvement of the Cloud service over its lifecycle must be shared between the provider and consumer, whereby the value of the service is defined by the consumer with the warranty assurance aspects supported predominantly by the provider. Refer to the earlier discussion (see "General Cloud Computing Framework Evaluation") regarding the consumer setting clear expectations with the provider on how service enhancements will reflect the consumer's priorities and needs.

- The provider and consumer should collaboratively apply the seven-step improvement process to define a superset of metrics that cover all process and technology metrics that one or both are concerned with.

- Service measurement and reporting management ensures that the defined metrics should be assigned some source and timing expectations, and then employed synergistically by provider and consumer in an agreed-on prescribed fashion to make improvements in the service.

B. Other Integrated Service Management Solution Implications

A key challenge here is associated with managing the Cloud's definition, design, and implementation processes where there is now the expectation for extensive virtualization of computing resources. The creation and application of highly reusable Cloud implementation and change components, templates and images will be especially important in making a quick, effective, and efficient Cloud.

It is one thing to develop and offer a comprehensive integrated service lifecycle management suite of solutions. It is another thing entirely to deliver this according to the segmentation of the target markets for the solution. It would be best to recognize the unique attributes of these markets, and perhaps deliver a spectrum of offerings that would fit the "demand package" needs of each market.

- The SMB (small and medium business) markets might be best served by one or more editions of a SaaS application with great interfaces that has a very strong but limited core of the leading standard features and functions.

- Mid-sized to large enterprises might prefer a SaaS solution that is more comprehensive, with greater depth and breadth than the SMB version but still retains the agility of SaaS delivery.

- Large, global companies may prefer a Super Cloud of SaaS public Cloud services and its private Cloud components, with the inherent portability of applications, tools and data to be able to move as desired between the two supporting Clouds. Portability will always be crucial to any consumer as its business needs change.

16

Service Offerings, Pricing, and Metering

Metrics and Charging Types

Discussion Areas in This Chapter

- *One-time*

- *Usage-based*

- *Advance versus arrears*

Service offerings, pricing, and metering

There is a broad spectrum of metered offerings, charge types, and pricing options available to the provider and the consumer. Note that these charges could be arms-length between independent provider and consumer enterprises, or charges allocated and distributed to internal enterprise Cloud service consumers.

A. Types of charges

Cloud Computing encompasses a variety of types of charges that service providers may use.

1. One-time charges

Even Cloud Computing providers may charge customers a one-time fee to initialize the service, to charge for other non-recurring implementation services, for administration services associated with major system changes.

2. Recurring charges

Recurring charges are basically period-based cycles of accruing usage information and billing for the period per established rates. The periods can be weekly, monthly, quarterly, or yearly. Charges can also reflect specified points when the charges are calculated, such as 1st or 15th of the month or a subscription anniversary date. There may also be specified proration rules that apply to partial periods.

There are also different types of recurring charges.

- Some are fixed as a flat fee that is charged each period.

- Some are variable calculated by multiplying purchase or usage units by the subscription rate.

3. Usage-based charges

Usage-based pricing provides the flexibility to charge for virtually any type of computing resource. Some commonly seen computing resources in the Cloud that need to be metered and measured on a recurring basis are:

- Per GB stored.

- Per IP address assigned.

- Per GB transferred, inbound and outbound.

- Per CPU instance, by size.

- Per virtual VPU instance, by size.

- Per OS instance.

- Per internet service.

4. Usage-in-arrears on-demand charges

Service providers typically make on-demand usage charges that are charged after the customer uses the service. This method is often preferred by the consumer because it offers the consumer the greatest flexibility with the least amount of advanced planning without long-term commitments. This additional flexibility for the consumer is charged at a higher rate and produces more revenue for the provider because this method makes it harder for the providers. They have greater difficulty to anticipate and plan for capacity changes, and it exposes the provider to consumers that switch to competing providers more easily and to the risk of bad debts after the resources have been consumed.

5. Paid-in-advance reservation charges

Service providers may allow consumers to pay in advance to reserve a specified amount of capacity. This method is generally preferred by consumers who have predictable, well-planned consumption requirements. Reservation pricing is generally offered at a lower price and produce less revenue for the service provider because they are able to better predict capacity needs, receive upfront payments to improve cash flow, remove bad debts issues regarding resources consumed, and reduce customer churn. The following are some examples of this method:

- Prepayment Plans: Consumers prepay a set amount for "use it or lose it" capacity, and may pay additional charges if that capacity is exceeded.

- Instance Reservation: Consumers reserve resource instances for a low one-time payment (e.g., for a 1- or 3-year term) and pay discounted usage rates throughout the duration of the term.

6. Location-based charges

A service provider may utilize multiple datacenters and may offer its consumers the opportunity to choose location preferences for certain workloads or for privacy reasons. The provider's datacenter costs may vary by location and it would pass on these cost-structure differences on to the consumer.

7. Peak and off-peak pricing

The service provider may have times when one or more of its resource capacities are close to full utilization or are underutilized, and it may choose to offer consumers subscription pricing plans with a standard, peak and off-peak set of rates to further use pricing to influence usage demand.

8. Free trials

Some SaaS application providers use free evaluation trials of their applications to reduce the consumer's perceived risk and make it easier and more likely for the consumer to convert to a paid subscription. These situations may also include:

- Requiring a credit card or other method of payment to be provided before the trial begins.

- Customization of the free trial duration period.

- Consumer notifications generated as the free trial is nearing expiration.

9. Promotions

Service providers may also use pricing to:

- Influence prospects and increase consumer acquisition

- Influence consumers to remain with the service provider.

- Move consumers from one pricing plan to another more profitable plan.

Providers may offer promotions that are:

- Time-based discounts offered for subscription by a date or a discount for the first month.

- A volume-based pricing plan representing a progressive tiered price or discount where increased volumes produce reduced costs to the consumer.

- A combination of time-based and volume-based pricing.

10. Service packaging and bundling

A service provider will influence consumer demand towards the services that it believes will provide the greatest value to its consumers at the best possible cost to the service provider. A very powerful way of doing this is accomplished when the provider defines packages and bundles that group services together to serve its capacity, service level and pricing objectives. If these packages and bundles accurately map to market segment demands, they will permit greater numbers of product configurations and incent quick access to consumer experimentation when any new plans become available to meet market or competitive changes as needed.

11. Resource allocation, pricing can be granular or general

Given the types of charges described above, service providers can offer their services in related clusters as packages with one price per period of time, or they can offer their services at very granulated rates that directly correspond to the allocation and usage of specific computing resources.

- SaaS applications may have packaged offerings as standard, premium and professional; or bronze, silver, and gold; or as novice, advanced, and expert. These packages may be a single subscription price per month, a price per month plus one or more additional usage-based charges, or just usage-based charges. A SaaS application's charges may encompass the entire stack of the application and the supporting cast of platform, infrastructure, security and management that is required to implement and run the application.

- Platforms offered as a service may also have packaged offerings and/or usage charges correlated to the number of objects, screens, tools used, API usage, library service calls used or added, etc. A PaaS offering's charges may encompass the entire stack of the platform and the supporting cast of infrastructure, security and management that are employed in developing, implementing, maintaining, and running the consumer's own applications.

Service Offerings, Pricing, and Metering

- Infrastructure charges can be very granular.

  o Storage and queuing charges may be driven by the amount of data stored, amount of data moved in and out of storage, and/or the number of requests made on the storage service.

  o Database service charges may be based on the amount of data that is stored, data transferred in and out of the DB, amount of time the DB instance is running, the amount of storage allocated to the instance, and/or the amount of CPU time consumed by query processing.

  o Compute service charges may be driven by the amount of time the instance runs, the amount of data transferred in and out, resource allocations made, storage volumes committed, etc.

17

Cloud Consumer Usage Considerations

Patterns, use cases, scenarios

Discussion Areas in This Chapter

- *Consumer startups*

- *Consumer moving to on-demand*

- *Software provider's operations*

Cloud Consumer Usage Considerations

The consumer's expected usage of Cloud Computing solutions should be well-defined when evaluating solution options. One part of this is in recognizing what the expected Cloud Computing usage patterns would be. Another part is identifying one or more application areas as targets for the adoption of Cloud Computing solutions. The remaining part is defining the options available for arriving at the desired Cloud Computing destination.

A. Consumer Usage Patterns

There are several distinct usage patterns of Cloud Computing resources that should be considered when evaluating prospective Cloud challenges and solution options.

- Usage that occurs at a fairly constant level with small deviations from a normal average is the easiest to forecast, plan, and support.

- Usage that has a highly predictable and well-substantiated steady growth rate with small deviations from a normal average growth rate is the next easiest to forecast, plan, and support.

- Internally-initiated loads

  o Cyclical usage that is driven by predictable periods of high usage, e.g. month-end, quarter-end, and year-end processes.

  o Exceptions, one-off usage bursts or spikes that are not easily predictable, e.g. large emergency processes or jobs

- Externally-initiated loads

o   Cyclical usage that has a high-positive correlation to events that are driven or closely supporting by web sites or web assets, e.g. concerts, professional sports games, etc.

o   Exceptions driven by unpredictable success or popularity of features on a web site, web casts, video, or new products, etc.

B. Cloud Consumer Use Cases

Cloud Computing has grown exponentially over the past several years and there are many reasons to expect that its greatest growth is still ahead. It is such a powerfully enabling and empowering paradigm that a tremendous amount of variation in Cloud Computing solutions is occurring. This is fundamentally driven by the fact that almost every aspect of the Cloud Computing framework can be offered as a service, and will likely find significant demand opportunities. There are some basic use cases for Cloud Computing solutions that should be established to make the overall evaluation process for strategy, planning, and executing more manageable.

Remember the various attributes of <u>Cloud Computing</u>? The consumer use cases will reflect the driving benefits of one or more of these attributes to define the reason for their presentation here as a foundational case.

1.   Use cases for initializing consumer startups

It is unlikely that any startup's investors or management team would endorse plans that are not based on extensive use of Cloud Computing resources to get the business up and running. A startup may go directly to the Cloud for most of its operations:

*   Implement most or all applications directly onto infrastructure resources that are delivering services from one or more external Cloud hosts. CRM, HR, Customer Self-Service, and eCommerce

platforms are examples of complete business functions that might first be exclusively operating within the Cloud, delivered over the Internet from external service providers as SaaS applications.

- Model the corporate network as Cloud resources in templates for instantiation as needed for:

  o Disaster recovery and business continuity situations.

  o Sandbox and staging for configuration, customization, load and compatibility testing, deploying new resources and upgrades with agility.

  o Conversion and assimilation of merged/acquired company's business data and processes into enterprise data and processes.

- The consumer could build their own applications on externally provided platforms. It could subscribe to a platform-as-a-service (PaaS, e.g. Force2.com) solution and create private enterprise Cloud applications to be delivered as services for internal consumer operations.

- The consumer could create, develop, implement, and maintain one or more internal Cloud platform(s), on legacy or Cloud infrastructure, that:

  o Support the development and deployment of internally-developed Cloud applications;

  o Support their integration with internal legacy applications, and external SaaS and PaaS solutions; and

  o Provide complete end-to-end business processes through services delivered to internal and partner consumers.

2. Consumers moving from on-premise to on-demand

These cases are presented from the perspective of a consumer that is moving to a mix of on-premise and on-demand. The chronology of the cases provided here reflects a risk-managing progression from on-premise to on-demand Cloud Computing solutions.

a) Basic Initial Consumer Cloud Cases for Migration from On-premise

Enterprises typically move to the Cloud initially in small steps that gradually provide increasing exposure to the moving parts of Cloud Computing while limiting the strategic and tactical risks to their business operations.

(1) Basic services

There are some business support resources that can easily be transferred from on-premise to on-demand. A good example of this would be the adoption of GMAIL as the SMS/email provider for a $50 per user subscription, with free access to other Google services, e.g. Enterprise Apps' Docs.

(2) Increase infrastructure capacity to meet increased operating demands

There are many moving parts within the IT infrastructure whose capacity can be increased in the Cloud to meet increases in the level of business demands.

- Host a static web site, with static HTML, CSS, and images content, through storage and content distribution services.

- Increase the storage capacity for scalability, long-term persistence, and economy.

- Increase compute capacity for processes, e.g. daily web logs, payrolls, daily reports.

- Increase the data processing capacity for unscheduled business and scientific analyses.

- Increase the capacity for processing and rendering of media files encompassing music, still images, videos, and large production projects.

- Increase a website's compute and storage capacity for (success) overflow.

(3)   Increase infrastructure capacity to add new operations

Expansion or upgrading infrastructure components might take months in the on-premise context, but may take minutes, hours, or a few days at most to add capacity to support new operations.

- Backup business data to archives in a Storage Cloud to support audit, compliance, and discovery requirements.

- Moving all data processing and analytics to the Cloud to provide separate application support resources for business operations versus reporting and analysis demands.

- Add one-off capacity for major compute and storage projects, such as conversion of hardcopy files to digital files and their distribution in various media.

- Add capacity for temporary storage and portability in moving data between datacenters, or in allowing comprehensive upgrades to on-premise resources.

- Add ETL (extract, transform, and load) capacity for operations in conversion from old systems and applications to new systems and applications.

b) Moderate Cloud Use Cases for Migration from On-premise

Based on their initial experiences with limited Cloud Computing solutions, consumers will begin to make more significant commitments to Cloud solutions. These are larger in impact but still reflect limitations in the scope and risk to mission-critical operations.

- Make extensive virtualization of infrastructure resources.

- Host the complex enterprise website in the Cloud; include legacy eCommerce applications and substantial server-side processing, database and storage, dynamic consumer presentation requirements.

- Consolidate the enterprise's globally-distributed datacenter operations into fewer (one to three) centers.

- Convert existing datacenter operations from on-premise to a private Cloud solution. Some people argue that private Clouds are not really part of "Cloud Computing". But a Cloud's resources should be virtualized, and this change alone is a catalyst for consolidation and a source of tremendous Cloud-driven cost reductions in IT operations.

- Add capacity through Virtual Private Cloud extensions.

- Move to support (ITIL) ITSM functions, processes, and lifecycle phases, through on-demand service management applications.

- Create hybrid web processes utilizing Cloud resources for some key business sub-processes.

- Subscribe to a platform-as-a-service (PaaS, e.g. Force2.com) and create private niche Cloud applications that are delivered as services for internal consumer operations.

- Develop internal platforms and then create private Cloud niche applications that are delivered as services for internal consumer operations.

c) Advanced Cloud Use Cases for Migration from On-premise

These are the most challenging cases for moving business operations to the Cloud.

- Model the existing corporate network as Cloud resources in templates for instantiation as needed for:

  o Disaster recovery and business continuity situations.

  o Sandbox and staging for configuration, customization, load and compatibility testing, deploying new resources and upgrades with agility.

  o Conversion and assimilation of merged/acquired company's business data and processes into enterprise data and processes.

- Move most or all legacy applications onto infrastructure resources that are delivered as services from one or more external Cloud hosts.

- Place entire business functions into the Cloud, delivered over the Internet from external service providers as SaaS applications. CRM, HR, Customer Self-Service, and eCommerce platforms are examples of complete business functions that might first be moved to be exclusively operating within the Cloud,

- Subscribe to a platform-as-a-service (PaaS, e.g. Force2.com) solution and create private enterprise Cloud applications to be delivered as services for internal consumer operations.

- Create, develop, implement and maintain one or more internal Cloud platform(s), on legacy or Cloud infrastructure, that:

  o Support the development and deployment of internally-developed Cloud applications;

  o Support their integration with internal legacy applications, and external SaaS and PaaS solutions; and

  o Provide complete end-to-end business processes through services delivered to internal and partner consumers.

3. Use cases for software provider's internal operations

Cloud Computing is a perfect solution model for software development throughout the development cycle and the software provider's associated value chain for going to market. The Cloud will ensure that all critical parties have sufficient and timely computing capacity for product strategy and planning, agile development, testing, training, demonstration, sales and marketing collateral production, post-sales consulting and professional services, and contiguous customer lifecycle support.

a) Strategy and planning

Product strategy and planning are extremely secure, content-rich and highly collaborative processes. Cloud Computing solutions could be used for all of these processes to meet these demands in days instead of months. Refer to Figure 16-1 below to see an overview example of the process that balances and aligns the strategies, plans and resources to lead to successful software development.

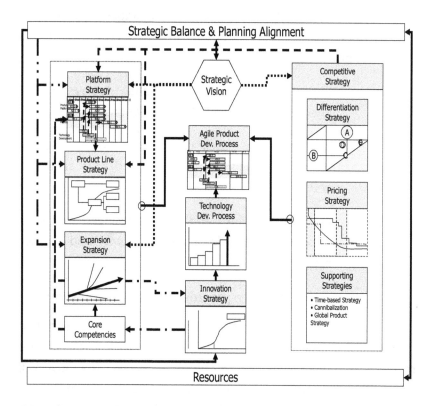

Figure 16-1: Flow and connections for software strategy and planning process.

The use cases would include:

- A collaborative executive-level process for the business unit's core strategic vision encompassing where the enterprise will compete, how it will compete, and the critical market, technology, and solution factors that will drive its success.

- A business unit-centric repository for market and competitive research, and the gathering, reviewing, valuing, and prioritizing of strategic product feature ideas and requests from all vested internal and external stakeholders.

- A collaborative decision support framework for defining, evaluating and aligning the core strategic vision with the critical strategies to ensure success. The alignment framework encompasses the business charter, financial strategy, innovation strategy, technology strategy, market strategy, core competencies strategy, product strategy, product line strategy, platform strategy, expansion strategy, differentiation strategy, pricing strategy, cannibalization and globalization strategies, etc.

- A collaborative planning solution that defines and maintains the enterprise's integrated financial, operational, sales and marketing plan, performance goals and metrics.

- A collaborative release planning and requirements management solution that supports agile development methods for marketing requirements, user stories, product and iteration backlog planning, estimates, use cases, non-functional specifications, iteration demos, iteration retrospectives, etc.

b) Development

Each developer must be working with their own unique environment of hardware and software platforms and tools. This is especially true when the Cloud solution is built on the three-tier architecture of web, application, and database servers. The Cloud as a development environment allows the developer to crank up all of the necessary servers, perform test-driven development as well as development-independent unit testing, and return the servers to the Cloud when they have completed unit tests, or reached a time or completion milestone.

c) Testing

Test-driven development in each developer's Cloud environment is the beginning of testing possibilities. Test plans can be run against a Cloud test environment where each source code commitment initiates a process of:

- Rebuilding the code line.

- Performing unit testing.

- Performing functional testing.

- Performing acceptance testing.

- Testing all supported platforms and environments.

- Testing load, scalability, and performance.

d) Training

The Cloud is a great solution for creating environments that are easily cranked up and easily shut down. Training is one of those function areas where:

- Cloud environments could be instantiated easily for each trainee.

- Training can be cumulative as well as highly reusable.

- Topics can be serially introduced or handled in parallel contexts.

- Topics in the Cloud can cover the spectrum of installation, configuration, and customization for preparation, as well as runtime processing and monitoring production environments.

- Training sessions can be accessed from anywhere at any time with full retention of the student's progress to date and timely feedback on ways to improve the value of the courseware.

e) Demonstration and Trials

Prospects want to see the value of the software provider's offerings before they will make any commitments to buy it. Standard software demonstrations usually provide about 70-80 % of the buyer's confidence with custom demonstrations and limited product trials providing the remaining 20-30%. If each sale engineer has to create their own demonstrations on their own portable machines, and rebuild these every time there is a system upgrade or a change in their machines, there is a lot of lost capacity to these overhead activities and a very inconsistent quality in the provider's performance.

Cloud environments could be constructed to provide pre-configured consumer solutions. The best sales solution engineers can create increasingly broad and deep demonstration scenarios, focused specifically on industry verticals and individual market segment needs, supported by scripts that address 90 to 100 % of the prospect's requirements. Additional environments can be cranked up from these demo environments to support pre-sales trials where the prospect can examine and play with the software to create their own tests, etc. This improves the quality of the sales that are made, increases the probability of a sale, tremendously reduces the sales cycle time and accelerates the timing of decision-making, reduces the cost of each sales cycle, and provides a very effective hand-off to post-sales and customer service functions.

f) Sales and Marketing

The marketing functions are extraordinarily enhanced by Cloud applications. Web-based marketing dramatically improves the efficacy of search engine marketing and optimization, while substantially reducing the cost to generate and acquire high-quality leads for sales activities. Cloud applications can provide these functions as well as provide integrated support for all other aspects of marketing, encompassing collaborative (with other internal functions and customers) creation of Market Plans, brand management, collateral production, success story production, event support, etc.

Product marketing participation in the definition of and access to the demo environments described above are also best provided through the Cloud's resources. And if the sales account executives are always using the optimal combination of Cloud-integrated provisioning of quality leads, marketing content and materials, demonstrations, trials, etc., they are fully prepared for success.

g) Professional services, Consulting, Customer Service /Support

The Cloud enables the software provider to support all of the post-sales activities necessary to bring the customer up to productive processes as quickly as possible and to maintain optimal customer performance and success in the use of their software. The hand-off of the presales configurations and customizations required to get to the sales commitment provides an initial leverage. But typically the customer's real production needs are not completely understood during the sale cycle, and professional services can use the Cloud environment to recognize opportunities to increase the effectiveness of the software, improve the consumer's business processes, and increase the probability for success in their business. The consultant can work with the customer to properly install, configure and customize their applications within sand-box and staging environments in the Cloud that assure implementation success.

Customer service and support can use the Cloud to create and retain an image of the customer's applications and provide fast and accurate solutions for any software challenge that may be causing problems for the consumer.

18

Software Provider Cases for Cloud Offerings

Startups and mature providers

Discussion Areas in This Chapter

- *New app and platform startups*

- *Enable installed base adoption*

- *Solution offerings' migration to on-demand*

Software Provider Cases for Cloud Offerings

The perspective presented here is that of a Cloud software provider that offers one or more Cloud Computing solutions as services. There are two sub-classes discussed here. In the first one, the software provider is exclusively offering Cloud on-demand solutions. In the second one, the provider must add one or more Cloud Computing offerings to its existing on-premise offerings.

A. Startup software provider use cases for on-demand offerings

A startup Cloud software provider can go to the market much quicker and with much less risk in initial investments (sunk costs) when it can start with a clean-slate to deliver Cloud services right from scratch. It doesn't have the challenge of balancing their offerings between the subscription and ownership license business models. Its architecture can be done "right" from the start, supporting service-oriented and multi-tenant applications.

The Cloud startup probably doesn't have an existing relationship with its own installed base to enable a faster adoption of its offerings. But it usually has far more agility in bringing offerings with substantial value to markets that have not been served very well by the competing on-premise vendors. And its offerings are likely to be far more attractive to the SMB markets than similar on-premise offerings.

One of the greatest challenges for the startup software provider is to focus its initial operations where they will have the most value in a balance between near-term tactical opportunities and long-term strategic goals. It is very rare that the business plan of a startup is an accurate reflection of what actually happens to the startup in the first 1-3 years. The value wasn't solely in the initial business plan, but primarily in the planning process that produced the business plan. If the startup can focus its initial offerings on the values that Cloud Computing will bring to its <u>targeted consumer markets</u> (see "Defining the Positive Attributes of the Cloud ") and to <u>its own operations</u> (see "Use cases for software provider's internal operations"), it will have the necessary agility that will probably be required to respond to the changes its markets will demand.

The case laid out below reflects the instance where the software provider's offering most closely adheres to the <u>Cloud Computing definition</u> (see "Cloud Computing Definition") that has been put forward in this book. It is arguably the most common startup model in practice now, and is only likely to increase as time-to-market and cost issues become more important to startups achieving success.

The startup could create its own data centers and all that goes with that. Unless there is some unique synergistic quality in its products, markets, or investors that will demand this, it should turn away from this option and choose a provider of its Cloud Computing framework stack for:

- The infrastructure that will optimally support the platforms and applications expected to be delivered as services to targeted markets, both in terms of design-time and runtime. The right flexibility in tools, libraries, services and service levels, with the desired elasticity and scalability to support success while avoiding disasters, with the right metrics at the right price and contract terms.

- The network services, foundation services, and Cloud management solutions that best fit the expected stack and targeted markets. If the startup is a PaaS provider, its search for fitting Cloud resources will likely end here.

- If the startup is to be a SaaS provider, it must search for the platform that best supports the immediate and long-term applications that are expected to be created and maintained in the provider's offering portfolio. The best solution might be to build a proprietary platform from scratch, if that is indeed the fastest way to get to the market with a great product. There is something to be said in this case for avoiding being "locked in" to a platform that will become a limiting factor or a liability over time. If the startup can see this to be the case, then the DIY (do-it-yourself) path may be best.

- But if the wannabe SaaS provider sees an external PaaS as the optimal solution:

  o Then there are other things to consider regarding the platform. If the startup chooses to build on an external platform offered as a PaaS, perhaps it can join an ecosystem that allows the startup to share its offering as one of over 1000 applications (e.g. the Force.com AppExchange). This would accelerate the startup's ability to get to the market and realize adoption quicker.

  o If the preferred PaaS has a roadmap that will probably cover all of the needs of the startup for the next 5 years, leaves it in a good position for moving to its own platform, and makes it a reasonable effort to move to any other platform, then this PaaS might be the best choice for a low-risk high-reward platform solution.

  o The startup might learn quite a bit about whether it makes best business sense to build its own platform, by using a PaaS solution for its initial foray into the Cloud platform business. Decisions informed by actual experiences are much more likely to be the right decisions.

- o Note that if the SaaS provider would move to its own platform, it could still retain its platform presence on the original PaaS for niche applications, thereby keeping its "hand" in that ecosystem while making its centerpiece platform-based offering solutions its own.

- Perhaps the SaaS provider will experience the type of core competency excellence that has occurred for Amazon, salesforce.com, and Google, and it will ultimately use this competency to create its own ecosystem of SaaS applications on its own PaaS offering.

B. Mature software provider moving to on-demand offerings

Software providers typically move from on-premise to <u>on-demand offerings</u> in incremental steps. The transition period's primary goal is to manage the business model risks associated with expanding the scope of offerings from exclusively on-premise to a balance of on-premise and on-demand solutions. The optimal solution will be if the subscription-based on-demand solutions have a nominal or acceptable level of cannibalization of the income from the on-premise offerings, but serve to expand to other segments within existing markets or address the needs of new markets where on-premise has not been nor will ever likely be successful.

The software provider must have a comprehensive vision and strategy that crosses both on-premise and on-demand offerings, and must consistently pursue a single, cohesive portfolio that will make sense to their business objectives over the long term, and make their markets and customers recognize the excellence of their vision, their management's ability to execute, and the leadership of their solutions.

1. Enable on-premise apps move to Cloud infrastructure

The initial step towards Cloud Computing would be to enable the installed base to move from their current original non-Cloud implementation of the applications to implementations that apply abstraction layers and virtualization of resources.

Many applications can be moved to the Cloud fairly unchanged, and the provider can work with leading customers and key technology partners to make this happen quickly. This change by the provider would remove constraints their on-premise only software had placed on their installed base consumers regarding datacenter consolidation, upgrading, etc. while allowing enterprises to implement a Private Cloud, extend to a Virtual Private Cloud, or move their infrastructure support entirely to an IaaS solution.

When competing with other providers for Cloud presence and new application sales, this provider can now effectively present the case that its applications are now running in the Cloud. The benefits to the consumer are significant though limited from taking full advantage of the Cloud's possible attributes.

2. Offer existing apps as SaaS multi-instance solutions

The software provider could offer their existing applications as on-demand solutions that are hosted by the provider in a Cloud-optimized, virtualized environment. Assuming the applications are web-enabled to start with, this initial SaaS offering could be built by simply supporting a multiple instance solution that resembles a multi-tenant solution from the consumer's perspective. The provider bears the incremental costs of the full single instance environment per consumer that it must support, but it can now provide an on-demand solution for those situations where a Cloud SaaS solution is required and there is no other current way to deliver it.

3. Offer existing apps as SaaS single instance solutions

This differs from the multiple instance solution above because it operates using a single instance of the application with limitations:

- Provides a UI that can be partitioned for each consumer's brand and skin

- Makes the applications "aware" of limited multi-tenant configuration abilities

- Provides a separate DB, on a single DB instance, for each consumer

- Separate load testing for each consumer

4. Create Cloud applications on partner Cloud platform(s)

The software provider's next step would be to use a partner's existing platform, subscribe to the PaaS as a consumer, create one or more applications on this platform, and make these applications available to consumers through a subscription-based SaaS model. The PaaS provider will likely have an ecosystem of applications that have been created on their platform.

This allows the software provider to:

- Create a SaaS Cloud presence offering multi-tenant SaaS applications for their brand and skin.

- Learn directly what the key tools and challenges are in constructing SaaS applications.

- Use an existing platform to determine what is required from a platform for their SaaS offerings.

- Initiate their Cloud applications strategically where they target the market segments with least disruption to their business plans and goals.

- Retain customers who were considering competing SaaS applications.

- Minimize competing SaaS application providers' ability to penetrate their markets / customers.

The consumer now has:

- An entry into Cloud Computing SaaS applications with a trusted vendor.

- Options to consider regarding the software provider's solution alternatives.

- Greater confidence that the provider will be a stronger partner in helping them to their optimal balance of on-premise and on-demand resources.

5. Develop or acquire their own Cloud application platform

The software provider will likely pursue the acquisition and/or development of their own platform in parallel with the above steps towards the Cloud. This is very critical for an on-premise applications provider if they wish to offer Cloud Computing SaaS applications and they want to have as much control as possible over how and when they do this.

The optimal solution is a new platform that is carefully architected to support ease-of-movement back and forth between the on-demand applications they build on this platform and their on-premise offerings.

- This ability enables their consumers to have great freedom in finding their balance between on-premise and on-demand at their own pace and in their own way.

- It allows the provider the means and opportunity to directly manage their Cloud business strategies and their portfolio without reliance on partners for strategic competencies.

The provider should create the platform to run on the most desirable Cloud infrastructure solutions, precluding solutions that their target markets will see as strategically limited, and driving those that fit the provider's business model goals.

6. Selectively create Cloud apps on its own Cloud platform

The software provider must use its own platform to develop and offer SaaS solutions that serve their target markets and segments in a manner that will minimize undesirable cannibalization of the on-premise revenues while providing the balanced value-adding portfolio of solutions. They may choose to enter new markets that are moving rapidly to the Cloud, focus on markets where competitive Cloud introductions are taking customers away, or target small and medium business segments within markets.

One of the first areas for consideration should be to offer solutions on their own platform that:

- Provide consumers with options for replacement and conversion from their analogous multi-instance and single-instance SaaS offerings.

- Minimize the customer's reliance on the applications built on the partner's platform while retaining presence in that ecosystem. Alternatively, they could replace and then improve upon those they offered on the 3rd party platform. When they do this, they should also introduce new applications or major new features that create additional value-based reasons for making the move.

The SaaS applications should have a fully multi-tenant architecture, supported by a multi-tenant platform and DB. This can be further refined if a layer for tenant load-balancing is provided to produce abstraction with the physical resources and even greater optimization of the load.

7. Offer the Cloud platform as PaaS with apps ecosystem

The software provider should build its platform to be offered as a PaaS solution supporting its own ecosystem of application software providers.

The benefits for this case include:

- Subscription income from the platform, consulting services, and professional services produce multiple new sources of Cloud revenue for the provider.

- It establishes the provider as a strategic partner to the consumer in their Cloud Computing strategies, plans, and execution.

- It allows faster and deeper adoption of the provider's balanced portfolio when all needs can be met through their offerings.

- It keeps the provider close to the market drivers and technology trends that are critical for effective continuous improvement in its own offerings.

- It provides direct insight into where its consumers see real SaaS application values.

- It provides the opportunity to support Cloud-based "networks" of consumers throughout their business processes, e.g. buyer and seller, manufacturer and reseller/distributor, etc.

- It allows the provider to argue that its portfolio provides everything that the consumer may need through its hopefully long and enduring lifecycle.

19

Waterfall and Agile Development Models

Comparisons and Cloud fit

Discussion Areas in This Chapter

- *Definitions*

- *Processes, roles*

- *Issues, solutions*

Waterfall and Agile Development Models

There are several common themes associated with Cloud Computing describing positive effects upon both providers and consumers, and this book's discussion of Cloud Computing has already provided detailed insights into many of their shared considerations.

Agility is one of the most prominent positive attributes that drives increasingly wide-spread Cloud Computing adoption wherever there are CPU's providing applications over networks to deliver valued results. In this context, agility is defined in this manner:

Optimal agility is the persistent ability to move to or deliver the best appropriate solution at the most opportune time and place, thereby consistently producing the greatest possible value in the results.

Cloud Computing will find it very difficult to achieve rapid, wide-spread adoption if their solutions have a limited capacity for agility. In fact, many might argue that it is the limitations that on-premise solutions have placed on both providers and consumers that have led to the success of the Cloud.

There should be agility in seeking, recognizing and responding to:

- Demand for new business solutions to retain or acquire a market leadership position.

- The critical trends in current or targeted expansion markets, as well as the requirements and priorities of customers and partners.

- The critical trends in enhancement, extension and disruptive technologies.

- Solution development, testing, training and implementation.

- Provider time-to-market cycles.

- Consumer time-to-productive cycles.

One particular area where this case is very clear is in the development of Cloud software platforms and applications. This chapter of the book provides a high-level comparison of the two predominant development methods used by software providers in most on-premise and on-demand software development projects. This book's high-level introduction to Cloud Computing includes the recognition that the best Cloud software solutions are likely to be developed in agile environments.

A. The Waterfall Development Model

What picture does the word "waterfall" conjure up in the mind's eye?

Well, depending on the waterfall examples that one might have witnessed, it might be a beautiful, long cascade of water falling in seemingly unstoppable volumes, overcoming the rock and plateau obstacles in its path, and arriving at a great lake or flowing river. Given nature's gravity, water falling is a very linear process, always falling downward. This is a very fitting and appropriate visual image for this software development model.

1. Definition

The waterfall model follows a traditional engineering cycle. It is a linear and sequential software development process that moves in a stepwise progression from requirements definition through development realization onto customer implementation and maintenance. There is typically a modest amount of overlap between the steps, and the overall period for a complete development cycle is usually between 6 and 24 months.

The linearity of its process requires a long cycle timeframe, and the length of the cycle produces a number of challenges for its agility.

2. Practices and roles

The overall process can span anywhere from 4 to 9 or more steps. Figure 19-1 below is a good representation of the flow and progression of the process. Each software release represents an iteration cycle of the Waterfall process.

The Waterfall Development Model

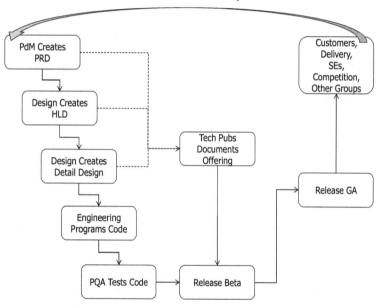

Figure 19-1 is a high-level view of the Waterfall model's process.

a) Product management provides requirements

In advance of this release cycle-centric step, product management should have developed a product roadmap that defines the strategic features of the finished product and assigns the realization of these features within one or more release cycles. The roadmap is the result of a company-wide product strategy and planning process that product management drives and manages.

With roadmap in hand, the Waterfall process focuses on one release. The initial step in the typical Waterfall process is the definition of requirements for the release. Product management conveys these requirements in detailed specification documents or PRD's (product requirement documents) to the engineering design, development and testing teams. The PRD is the document that drives everything down-waterfall, so if it isn't in the PRD, it shouldn't be in the release.

b) Design team creates a high-level design

The first "plateau" for the Waterfall process follows the product management PRD as it is received by the engineering design team. Unfortunately, in the course of its work over a period of weeks or months, the design team determines questions and issues across the scope of the entire release that the original PRD doesn't address or resolve. They take these questions back to product management, and in the course of these discussions, the answers to the Design questions are sufficient to allow them to create a HLD or high-level design document. Product management will likely revise its PRD to reflect whatever requirements issues that were raised and resolved with the design team.

Note that this is the first of many loops in the process, as depicted in Figure 19-2 below.

The Loops Within the Serial Waterfall

Figure 19-2 depicts the process loops of the serial Waterfall process

c) Design team creates a detailed design

The design team has a HLD that reflects the current revised PRD. It takes this document down to the next level of design, creating detailed specification that the development team will use to create the code and overall product features, etc. Unfortunately, in the course of this detail work over a period of weeks or months, the design team determines new questions and issues across the scope of the entire release that the revised PRD doesn't address or resolve. They take these questions back to the HLD authors and product management, and in the course of these discussions, the answers to the detail design questions are sufficient to allow them to create a detail design document. Product management will likely revise its PRD and the high-level design team will likely revise the HLD wherever necessary to reflect the requirements issues that were raised and resolved with the design team.

d) The developers create the code and solutions

One should be able to recognize a pattern forming here that is endemic to the Waterfall's lengthy linear process cycle. The entire scope of the release is passed through a progression of "throwing it over the wall" milestone points that take place with each step.

In this step, the design team's detail specifications have been formally provided to the engineering team as the Detail Design Document. The DDD reflects the current revised PRD and HLD. The engineering team will use the DDD to create the code and overall product features, etc. Unfortunately, in the course of this development work over a period of months, the engineering team determines new questions and issues across the scope of the entire release that the revised PRD, HLD and DDD don't address or resolve. They take these questions back to the previous authors, and in the course of these discussions, the answers to the questions are sufficient to allow them to create the code for the solution. All up-waterfall documents will likely be revised to reflect the requirements issues that were raised and resolved with the engineering team.

e) Product quality assurance tests the code

Quality assurance will test the code delivered by engineering as complete and ready for various acceptance testing activities. The test cases should reflect the expectations expressed in the up-waterfall documents. The engineering team may have done some unit and acceptance testing already, but the primary testing burden typically falls on QA in a subsequent step in the Waterfall process. During the course of this test work over a period of weeks or months, the testing team finds "bugs" where the software doesn't behave as expected according to the PRD, HLD, and DDD. The test team returns this to the engineering team for fixing, until it gets it right or until the up-waterfall documents change to be in synch with the engineering team's actual code solution. All up-waterfall documents will likely be revised to reflect the requirements change issues that were raised and resolved with the test and engineering teams.

f) Technical publications documents the release

Users will require some information for the entire scope of the release regarding what the new/changed/deprecated software features are, how they work, and how the users should best make them work as desired. During the course of this documentation work over a period of weeks or months, the documentation team finds "issues" where the documentation content that reflects the software's behavior is not likely to make sense to the users. The code may need to change and/or all up-waterfall documents may need to be revised to reflect the change issues that were raised and resolved.

g) Release

The software release has been under construction for several months at this point. The release management team will likely try to have a very limited Beta or CA (Controlled Availability) release for a few select customers in advance of the general software release, with the intention of finding and fixing any bugs or issues that were not caught by the provider's internal processes. During the CA or early general release timeframes, the provider will have to deal with these issues, and this may again require changes in all levels of documentation, testing, and processing.

h) Feedback closure

There are many stakeholders associated with the success of the software. The provider's sales, marketing, consulting, and training functions are directly affected by the release. As these functions have increased exposure to the release and the market's response to the release, they develop feedback for the product management and other Waterfall process functions. This feedback may be positive, negative, or feature requests driven by market changes and competition. This content is considered as minor fix-it releases occur, but new features are typically the exclusive domain of the next major release cycle.

3. Issues with Cloud

If agility is important to the Cloud's providers and consumers, there are several issues associated with the Waterfall model that would make it a less desirable model for Cloud platform and application software development.

a) The frequency of the iteration cycle

The release cycle may range between 6 and 24 months, with the average being around 15 months. The scope is basically set when the product management roadmap is driven down into detailed requirements expressed in the PRD, and every step after this is simply trying to address the entire release scope. The consumer must wait for a very long period of time to experience any of the expected functional, security, or productivity gains, keeping its business processes static.

b) Passing the entire release scope across each step is inefficient

The Waterfall process requires the entire scope of the release be passed from one sub-process owner to another. This requires each sub-process owner to deal with every detailed aspect of the entire release until all details have been satisfactorily resolved. The lack of priorities that reflect the value-based demands of the market means that everyone along the Waterfall path spends energy on every aspect of the release scope, making the really important features wait as long as it takes to solve the most unimportant challenges. Since all of the scope must be resolved before the current step is completed, the release cycle becomes a long one.

c) The value of the features is diminished

The features and/or the behavior of the features that are finally delivered in a release may not represent what the targeted consuming markets expected. It is also quite possible that the consumers have encountered new business challenges and/or changing market priorities that, even if the released feature scope and the related behavior is what was expected at the start of the release's Waterfall process, the software cannot meet the current or new-priority needs of the consumer.

With the Waterfall model, there is no wiggle room for error recovery. Imagine consuming more than a year's worth of capital, resources, etc. to deliver a solution that doesn't appeal to the market while the more agile competitors may have delivered several productive consumer-satisfying installments of improved productive software over the same period.

d) Sub-optimal results

Every sub-process owner is serially engaged to swallow and digest the entire "elephant" (release scope), meaning that the up-waterfall sub-process owners must be available to assist each team in their digestion. This makes for a very unconstructive and counter-productive way of producing software, because it is more likely to hide inefficiencies, permit bottlenecks to go unresolved too long, and conflicts between the various teams, which in turn dramatically increases the politics around ownership, transparency, responsibility and accountability, and places enormous pressure on providing the right requirements for the entire release scope nearly 2 years in advance of the release date.

From a risk management perspective, Waterfall involves a long cycle that conditions its stakeholders to consistently reduce expectations and accept compromised results.

One or more things are likely to occur when products are consistently released with less quality, higher costs, longer time-to-market, and increased cost to consumers from receiving less than expected, later than expected. None of them are good, so it may feel like "picking your poison":

- Move the release date out, providing more time for the release to meet expectations in scope and quality without increasing resources.

- Add resources to meet expectations on release date, scope, and quality.

- Reduce scope while meeting expectations on release date and quality with the current resource levels.

- Reduce quality assurance to meet the expected release date with the expected scope and current resource levels.

- Move the release date out, add resources, reduce scope, and reduce quality (Yikes, the perfect storm!).

B. Agile Scrum

One would be correct assuming a development model that actually had the word "agile' in its name might in fact be more agile than the Waterfall model of software development. Agile methods are defined here while Scrum is defined as a particular agile method on the following page.

Agile software development methods can be traced back to Lean Manufacturing and Six Sigma concepts that emphasize error proofing, elimination of waste, creating optimal process flows, adding value to the customer, and empowering workers. While embracing these roots, agile software development recognizes that it is focused on software.

These methods employ iterative and incremental development cycles that emphasize the collaborative work of self-managing cross-functional teams to produce frequently-improving market-driven software solutions. A "Manifesto for Agile Software Development" was created and published in 2001 that reflects the approach of these methods (bold and underline added for demarcation emphasis):

"We are uncovering better ways of developing software by doing it and helping others do it. Through this work we have come to value:

> ➢ Individuals and interactions **over** processes and tools

> ➢ Working software **over** comprehensive documentation

> ➢ Customer collaboration **over** contract negotiation

> ➢ Responding to change **over** following a plan

That is, while there is value in the items on the right, we value the items on the left more."

The manifesto led to "lightweight" software development methods being described as agile methods, including Scrum, Crystal Clear, Extreme Programming (XP), Adaptive Software Development (ASD), Feature Driven Development (FDD), and Dynamic Systems Development Method (DSDM).

Scrum is one of the more predominant methods in terms of usage, but most agile methods are in practice an adaptation or mix of methods. This book will look only at Scrum specifically in making this comparison to the Waterfall method.

1. Definition

Scrum is a framework model for building complex solutions through optimal processes and techniques. It provides roles, time-boxed activities, and process artifacts that enable an integrated iterative process of designing, building, testing, and delivering solutions. The fundamental process revolves around the formation of self-managing teams that take small steps to deliver manageable increments of working software, while empirically adjusting their process to increase development planning accuracy, optimize product quality, and minimize risks.

Quite often, the difference between success and something less than success are the decisions one makes around tradeoffs regarding what to emphasize. There are several key attributes that are at the center of what really defines the Scrum model. Using the outline provided by the Agile Manifesto above, some of these key attributes are:

a) Customer collaboration is valued over contract negotiations

It is better to have a consumer that interacts with a provider as a close partner in achieving success than it is to operate with each exclusively at arm's length using the T's and C's of a contract as the relationship roadmap. A solution that serves no market will never succeed as well as a solution that best meets a market's critical need.

- The service consumer will always be most satisfied by the consistently rapid delivery of high-utilization software that best meets their needs according to their priorities and expectations. And a very satisfied consumer is the best kind of partner to a provider.

- Given the above definition, the most effective way for the provider to ensure very satisfied consumers is through frequent, timely and close communications between its development team(s) and the targeted consumer base(s).

b) Working software is much more important than comprehensive documentation

When a software solution actually <u>works intuitively</u> (see "Core application technologies") in the most useful manner to deliver the functions that are needed by the consumer, the burden on documentation should logically be substantially reduced. And if one can create more of the software that fits this definition rather than extensively documenting that it works, let there be more consumer happiness over delivered productivity and greater evidence that the chosen provider was the right choice.

- Reducing the cycle of delivering working software down to shorter periods with frequent deliveries means that every participant involved in the process – providers and consumers alike – receive much more timely and constructive feedback. This means that the risks associated with longer development cycles are largely mitigated if not completely removed, and that the value of what is delivered is much greater for all parties.

- The best indicator or measurement that the provider's software development resources are making the right progress and are being optimally applied is frequent deliveries of working, useful software.

- For finely tuned processes, there is usually a best rhythm that will produce the desired results. Software development that has such a rhythm or pace is much more likely to be able to sustain expected or predictable development success levels.

- The over-engineered or complicated solution that allows a 98% solution fit is probably not nearly as effective from a cost, utility, or time-to-market perspective as the simple solution that has an 85% coverage rate. Providers that delay delivery of a solution as they pursue a "best" initial solution should recognize that the provider that first meets the consumer's real values and needs is the winner, whether as a pioneer or a close follower to the market.

c) **It is better to change when necessary than to continue according to the original plan**

One of the most popular perspectives shared regarding plans and planning comes from Dwight Eisenhower, whose great administrative skills were a key reason for his appointment as overall Allied commander in World War II Europe. He stated, "I have always found that plans are useless, but planning is indispensable." Many people think that planning efforts are less with Scrum, when the reality is that Scrum requires more planning effort than a waterfall model. Because when "the thing" that originally drove the plan has changed, and the process is empirically-driven, the plan must change as well.

- The provider must understand that developing and delivering software solutions that do not meet the current and future needs of its targeted consumers is the absolute worst thing for everyone involved. The working capital is wasted, there will be no return on the provider's investment, the time has been lost and competitors who have delivered the solutions that do meet the current and future needs of the targeted consumer have a differentiable competitive advantage now. So responding to changing requirements, even late in the development process (another advantage of short cycles), is preferable to the alternatives.

- The process needs to have a cycle component where careful scrutiny of the most recent actual performance is expected to surface recognized issues and produce the necessary enhancements to meet changing circumstances. This ability to adapt applies to both the process and the solutions produced by that process, and the key is that the process provide steps, activities, or formal opportunities for this adaptability to take place.

- Scrum is empirically-driven. This means that there will always be continuous attention given to sustaining good design and technical excellence. The provider understands that a poor or even mediocre design will produce sub-optimal results and probably require re-engineering and/or rework, and the consumer is likely to be negatively affected. Further, technical excellence means leadership in providing the best technology that the consumer can find.

The Deming Cycle can be a good representation of the empirical basis for Scrum or other agile methods within the Sprint cycle. See a graphic of this cycle below in Figure 19-3.

Empirical Process Improvement

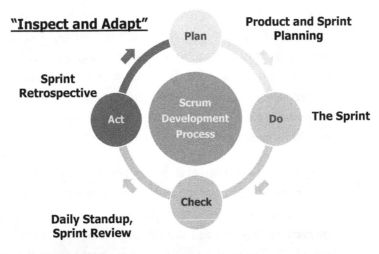

Figure 19-3 Deming Cycle: Continuous improvement of processes.

d) Empowered people will work together in spite of limitations in process or tools

Prescribed processes and designed tools are the result of people figuring out how to do things in a better way, and the process of finding ways to do things better is continual.

- The best people to have working on a project are those that are highly-motivated to produce project success for the preferred reasons. Most managers will recognize that people who either lack motivation or whose motivation is fueled by the wrong drivers will require too much direction, destroy the project chemistry, or fail at the worst possible time for the project's success. Assuming that someone is properly motivated and that they are capable of delivering the required work, then they absolutely must be trusted to deliver the work. Absence of trust will substantially erode their confidence and destroy their motivations.

- One of the greatest forms of trust is evidenced by allowing a team to be self-organizing, self-regulating, and self-managing. It is management 101 to recognize that there must be truth and honesty within a team, and that this transparency is only meaningful where there is the acceptance of individual and team responsibility, and the expectation of peer-reinforced accountability.

- Technology is great, and there are so many ways to communicate. But the most effective and efficient way to communicate is still and likely will always be through physical co-location and face-to-face dialogue. One of the key challenges to the provider is how best to pull this off within a global network of development resources.

2. Practices and roles

Scrum has some of the same roles as the Waterfall model, albeit with different deliverables and context, as well as some new roles. A comprehensive presentation of the Scrum processes would require substantially more than what is presented here. This book provides a basic overview of the flow, and the key practices and roles of the process to enable its comparison with the Waterfall model, and its positioning as the best development model for the Cloud software development activities.

a) Product backlog

No surprise here. The process assumes that the provider's product management team represents the business needs of the targeted consumers. As was the case with the waterfall model, product management should have developed a product roadmap that defines the strategic features of the finished product and assigns the realization of these features within one or more release cycles. The roadmap is still the result of a company-wide product strategy and planning process that product management drives and manages. Figure 19-4 below provides a graphical illustration of the considerations that enter the product strategy funnel and the basic components of the strategy-setting sub-process.

Strategy and Planning Process
Sprint-based Release Time-to-Market Cycle

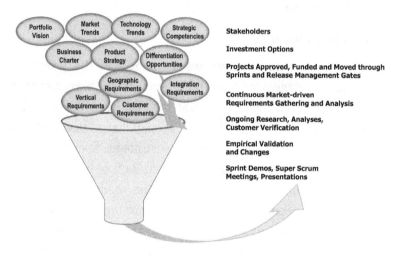

Stakeholders

Investment Options

Projects Approved, Funded and Moved through Sprints and Release Management Gates

Continuous Market-driven Requirements Gathering and Analysis

Ongoing Research, Analyses, Customer Verification

Empirical Validation and Changes

Sprint Demos, Super Scrum Meetings, Presentations

Figure 19 – 5: A strategic process drives the product roadmap

b) Release planning

This is the process that first divides the product roadmap feature sets into assigned releases. The result should be that the product management team creates and maintains a comprehensive Product Backlog, hopefully beyond the next release but the next release scope is required. The Product Backlog consists of individual user stories for each key element or unit of features to be developed. A user story typically consists of 3 parts: The user focus, the desired action or activity, and the expected result or benefit.

One example of a story might be:

"As a product owner, I want to be able to create and post user stories within the Product Backlog, to establish the high-level

feature expectations for the end-state of the software product and enable the development process to move forward."

Individual members of the product management team are designated as Product Owner for some clearly isolable features, feature sets, product lines, etc. The Product Owner, as the voice of the consumer, is the role that creates the associated product backlog for their assigned scope. Further, the product owner orders the Product Backlog according to the business value and priorities that best map to the consumer's needs and expectations.

Scrum development is done in iterative cycles typically referred to as Iterations or Sprints. A deliverable of the Release Planning process is the assignment of the release's Product Backlog items to individual Sprints or Iterations cycles. Note that where extensive research is in the critical path to proper item definition, estimation, and planning, an initial research "Iteration Zero" should be used whose results will feed the following (now more definable) iterations.

c) Sprint planning

As user stories get higher in the Product backlog queue and closer to the Sprint that they will be assigned to, they are increasingly groomed to provide sufficient granularity of the requirements for accurate estimates and sizing of the required development efforts. It should be noted that backlog grooming continues even during the Sprint itself, but most of the effort during a Sprint is focused on grooming the backlog for the next Sprint, not the current Sprint. The Scrum Team can be from 5 to 10 members, and consists of the Product Owner, a Scrum Master to facilitate the Scrum process activities to ensure Scrum disciplines are followed and enable

Scrum success, and a development team consisting of the developers, and, as warranted by the Sprint scope, architects, UX designers, QA, writers, etc.

The Sprint planning process covers 2 steps.

- The first step focuses on what should be built in the Sprint being planned. The Product Owner presents the product vision and the results from the last iteration to provide big-picture context, and then identifies the requirements that should be assigned from Product Backlog into the Sprint backlog. Fit here considers the minimal functionality of feature sets required to be marketable, the velocity and capacity of the development team in addition to the priority. The final deliverable here is a micro-charter or Sprint goal definition that briefly but clearly establishes the objectives for the Sprint.

- The second step has the Scrum Master and the development team meeting to establish how they plan to build the Sprint Backlog items. They effectively break the items down into tasks that are expected to be required in building working software. They express sizing estimates by using relational "points" rather than time period unit estimates.

- As a result of the Sprint Planning meetings, items have established business value expectations, clear user stories, granular well-defined requirements, and acceptance criteria for completion. The items that have been accepted into the Sprint Backlog represent a commitment by the development team members to deliver the included items to the Product Owner as working software at the close of the Sprint. There will be no external influences allowed to affect the development team during the iteration. The definition of working software is clearly established as well, putting all parties of the Scrum team on the same definition regarding what "DONE" means. Refer to Figure 19-5 below to see the overview of the Sprint's key moving parts.

The Sprint Processes

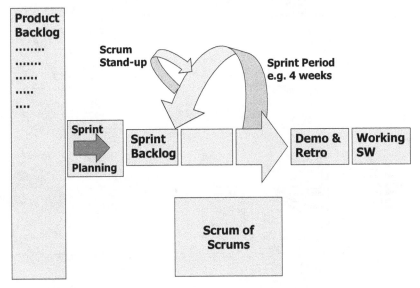

Figure 19-5 represents the Sprint process flow and components.

The development team creates a board and some charts that are prominently placed in the working area of the team.

- A **Task Board** is created and maintained for the Sprint, that covers the movement of requirements-driven development tasks through status of "to do", "assigned", "in process", " to be verified", and "done". Done status means that the task has been verified and reported within the daily standup meeting (see below) to meet the criteria of "Done". Figure 19-6 represents an example of the large physical board that exists in the development team's work area. Each of the blocks in this graphic represents a 3x5 card or some other small artifact that is placed on the board and then moved as appropriate.

Example of a Sprint Task Board

Sprint Backlog	Task Breakdown	Tasks in Progress	Work to Verify	Done
Story: 1: 5 points As a user, I want to ..., to provide ...	Code for ... / Code for ... / Test for ... / Code for ... / Test for ...	Code for ... / Test for ...	Test for ...	Code for ... / Test for ... / Code for ... / Code for ... / Test for ...
Story: 2: 9 points As a manager, I want to ..., to ensure ...	Code for ... / Test for ... / Code for ... / Test for ...	Code for ...	Test for ...	

Figure 19-6 is an example of the Sprint Task Board.

- Burn-down and burn-up charts are two different ways of looking at the progress that the team is making against the Sprint plan, using the story points on one axis and the Sprint's days on the other axis. One can quickly use the simple charts to assess the team's positional status relative to the Sprint plan, and the team's velocity (pace or points per day).

For example, the burn-down chart in Figure 19-7 is reflecting the progress that the Scrum team is making towards its committed scope of deliverables for the Sprint. See figure 19-8 for an example of the burn-up chart, which might provide a better big-picture view that the team's velocity in delivering expected scope is less than originally committed for the Sprint. These charts are typically maintained daily, and then placed on the Task Board, usually under the "in process" heading

- Note that there are also several other similar charts that deal with monitoring the velocity (pace or points per day), or showing the progress across multiple iterations in a Release Burn-Down Chart, etc.

Example of a Sprint Burn-down Chart

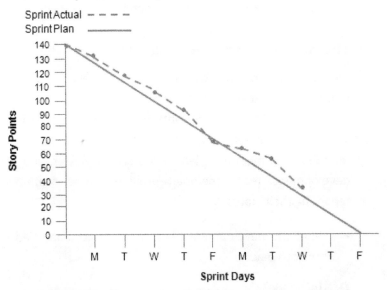

Figure 19-7 is an example of a Sprint burn-down chart.

Example of a Sprint Burn-up Chart

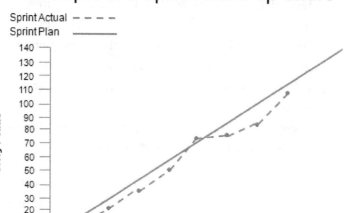

Figure 19-8 is an example of a Sprint burn-up chart.

d) The Daily Scrum or Sprint Stand-up

The members of the Scrum Team meet every day for a meeting of approximately 15 minutes. The Product Owner is optional but all other participants are required. All participants stand for the meeting (a great incentive to be done in 15 minutes). The meeting is led by the Scrum Master, and each development team member provides three statements regarding their status:

- What the development team member accomplished the previous day.

- What the development team member expects to accomplish before the next day's standup.

- What are the obstacles that the development team member is facing that may affect their ability to meet the expected accomplishments.

This is about communication, transparency, self-regulation, and maintaining trust by delivering. The team looks at the charts as well, and the Scrum Master may call for follow-on meetings if they are needed to help the team stay on track with the Sprint expectations or remove some obstacles.

e) Scrum of Scrums and Stakeholders Meetings

There may be multiple Scrum teams working on various areas of the software offering, perhaps with direct interdependencies, such as one or more platform teams working on the platform while one or more application teams are working on the applications that use the platform. When there are globally distributed Scrum teams, a Scrum Master at the remote site works with a Scrum master at the centralized site to facilitate daily local Scrum meetings, etc.

There may be 2-3 meetings per week then that are Scrum of the Scrums meetings, where the Scrum master or other qualified member of each Scrum team represents their team in a Scrum type of "stand-up" meeting.

There may also be a stakeholder's meeting once every couple of weeks, whereby the product owners provide Sprint and Release status information to the stakeholders that will be affected by the success of the expected Sprint. These discussions or follow-on activities are not allowed to distract or interfere with the teams' progress against the current Sprint.

f) The Sprint Cycle

The duration of the Sprint is set at planning. Some providers are very aggressive and set the Sprint as one week, say Tuesday through the following Monday. The most common Sprint period is a month or 4 weeks, and the very longest Sprint cycle is maybe 4 months. With the more lengthy periods, much of the "goodness" of iterative software development is likely to be lost.

There are 2 critical Sprint meetings that occur at the end of the Sprint.

- The Sprint Review or "demo" meeting provides a demonstration (no slides) of the working software to the stakeholders (targeted users, management, consumer representatives, partners, etc.). The Product Owner and/or Scrum Master work to make the constituents aware of what has been built compared to the expectations and goal for the Sprint. The feedback and status feeds right into the next Sprint Planning meeting.

- The Sprint Retrospective follows the demo meeting. The Product Owner may or may not be invited to participate. The Scrum master leads the meeting with the development team to discuss what went right and what went wrong in the just-completed Sprint and what might be done to change the process as needed to improve the outcome, etc. This is the one of the most visible aspects of the empirical and adaptive nature of Scrum. Figure 19-9 reflects all of the circles where the Scrum process may provide feedback and improvement to the process.

Possible Scrum Feedback Cycles

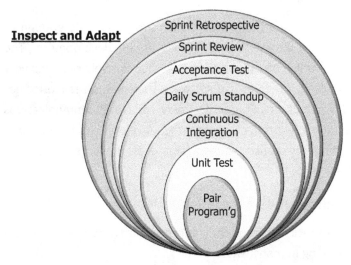

Inspect and Adapt

Sprint Retrospective

Sprint Review

Acceptance Test

Daily Scrum Standup

Continuous Integration

Unit Test

Pair Program'g

Figure 19-9: Feedback cycle opportunities the Scrum model might provide.

3. Scrum is the perfect fit for the Cloud

The Scrum model for software development is a much better fit for Cloud Computing platform and application solution providers than the Waterfall model alternative. No wonder many of the leading Cloud platform and applications developers employ Scrum throughout their development processes. One leading example of this is that sales force.com and Force.com are highly proficient in their complete reliance upon very refined Scrum processes, and they have been held up as a great example of its major contribution to Cloud providers and consumers alike. Not every Cloud software provider could have pulled off the cold-turkey "big bang" implementation of Scrum as they have, but their Scrum process has become a core competency, a critical differentiator, and a primary factor for their ongoing "rhythm" of success.

Consider the following conclusions that support Scrum as the more agile of these compared methods:

a) Close customer collaboration is achieved

Scrum development processes are small and agile enough to be heavily biased toward market-driven development. The consumer will be quite satisfied by the relatively rapid delivery of working, ready-to-use software that meets their priorities and expectations. Frequent, timely and close communications between the product owner, the development team(s) and the targeted consumer base(s) are at the core of Scrum's value.

b) Working software is achieved

Scrum's focus on frequently delivered software that works is a good thing in many ways.

- Frequent deliveries of working software means that every participant involved in the process – providers and consumers alike – receive much more timely and constructive feedback. Therefore, the risks associated with development cycles are largely removed.

- The Sprint iteration cycles make everything adopt an optimal rhythm that is able to sustain expected or predictable development success levels.

- The provider has a much higher ROI sooner, and the consumer achieves a broad increase in productivity and business operations success sooner as well.

c) Multiple points for feedback assures best results and continual improvement

Scrum is a highly-adaptive process, with feedback cycles from multiple points, including test-driven development, daily stand-up meetings, performance charts, Sprint demos and retrospectives.

- The provider is developing and delivering software solutions that meet the needs and priorities of its service consumers. Scrum applies the just-in-time principle to every aspect of the process, assuring that waste has been substantially reduced if not eliminated. There is no working capital wasted, faster access to revenues from earlier releases of working software, no missed-opportunity losses to competitors, and continual leadership.

- Scrum's adaptability applies to the process and the software solutions from the steps, activities, or formal opportunities where this adaptability is supported.

- Scrum provides continuous attention to sustaining good design and technical excellence. The provider's emphasis on design and technical excellence means leadership in providing the best possible levels of utility and technology.

d) Teams of motivated and trusted people will deliver results

Scrum relies completely on a team concept that is development-friendly at its core. Development leadership determines what demand solutions they will commit to in a Sprint, what resources by type and number they will require to be successful, and the day-to-day ability to self-organize and self-regulate their team efforts.

Trust is best earned and rewarded based on performance meeting or exceeding expectations, often in spite of other influences. But trust is so intrinsically woven into the fabric of Scrum Teams that Scrum would likely fail without it. The development team determines the expected workload of the items it commits to in a Sprint, the expected development team velocity, the expected development team capacity, and only then, having fully determined what it will commit to, OWNS the responsibility and accountability to meet the Sprint's challenges.

Scrum teams support a Sprint cycle that is typically defined in weeks rather than the Waterfall's years. Given Scrum's collaborative, non-linear process that emphasizes market-driven priorities, there is a feeling of "us" working together rather than the "us versus them" of a serial throw-it-over-the-wall approach.

There is room for change, but the more iterations that a team goes through working with each other on a platform and/or application, the higher the credibility and accuracy of their planning and estimation efforts. At some point in their Scrum process maturity, they will consistently deliver the expected or better scope and quality at the expected dates and with the expected resources. And this is a great recipe for agility and success for Cloud software service providers and their fortunate Cloud service consumers.

Thank you for your time and interest!

Good luck with your Cloud shapes!

Closure (aka The Celebration)

Index

Index

Index

Index

Index

www.ingramcontent.com/pod-product-compliance
Lightning Source LLC
Chambersburg PA
CBHW071407050326
40689CB00010B/1780